"Khaki"

on cotton and other textile material.

By

Dr. Friedrich Carl Theis.

(Translated by E. C. Kayser.)

With 64 specially prepared patterns.

Berlin W.

Published by M. Krayn.

London.
Heywood & Co. Ltd.

1903

New YORK.
G. E. Stechert,
9 East 16th St.

Authors preface.

A monograph, devoted to one particular mode-shade, would appear somewhat risky, had not preceeding events in so distinct a manner paved the way, as with "khaki".

The Transvaal war and the disturbances in China, crowded together into a short space of time, unexpectedly and suddenly called upon the dyer for a shade, whereof the correct manner of production had yet to be studied and essayed.

In this experimental stage, the right key could not always be struck at once; much inferior work was also turned out, to secure enhanced profits.

War and hurry being over, the time has arrived for sifting and discrimination.

Recent campaigns have tought the necessity of adapting the color of army clothing to that of the soil; on that account, "khaki", in its various modifications, may well be credited with having before it a prolonged existence.

Hence this pamphlet, the purpose whereof is, to record sound facts and to remove errors and secrecy.

This aim could not be achieved satisfactorily, without paying due attention, besides to true metallic-oxide-khaki, to the various imitations, produced with artificial dyestuffs, the so-called "sulphur-colors" in particular. If such dyeings do not attain the solidity to light and exposure, which distinguishes metallic oxide colorings, they have, on the other hand, so many practical advantages, that, from a general point of view, one may well say: "For clothing — artificial dyes; for tents — metallic oxides."

To the various colorworks, that have in this direction assisted me with patterns and recipes, I herewith beg to offer my warmest thanks.

Experts may perhaps, at the first glance, be inclined to see in this publication but a promiscuous compilation. It was, in fact, no easy matter, to connect, in anything like logical manner, the various methods and patents; if it be conceeded that this task has been solved to some extent at least, such a verdict will give ample satisfaction to

Ohligs, autumn 1902.

The Author.

Translators preface.

The translator, whilst be sees no reason to conceal his identity, would not arrogate to himself a tittle of the credit, due to the author for, what no doubt will be judged, a painstaking and laborious piece of work.

Nor would he, on the other hand, be deemed on every point in accord with the opinions of the author or the indirect utterances of others, expressed in these pages.

Declining all criticism, he has merely restricted himself to his nominal duties, and such omissions, curtailments and alterations, as he has ventured upon, were made merely, in order to remove, what to the English reader might appear tedious, superfluous or perplexing.

The authors preface, which must apply to the translation with almost equal force as to the original, fully explains the cause and purpose of this publication.

Perhaps it may be said with some show of reason, that, to import information on khaki-dyeing, into England, is "carrying coals to Newcastle". The larger opportunities, which prevailed in this country for gaining experience in that particular line, may have revealed facts, still little familiar or even unknown to the continental dyer. If that be so, this superior knowledge is restricted to few and has not yet found utterance, and whilst some may think they could, if they would, have written a more up-to-date account of khaki-dyeing, a great many more will undoubtedly be glad of the information, conveyed to them in these pages.

That khaki goods are at the present moment, and have been for some time, a drug in the market, is known to everybody, nor is there much probability of the article ever again coming into "fashion". In compensation, there is the certainty of a steady and running demand for army-purposes. Competition for these orders will, no doubt, become keener every year, whilst deliveries will have to stand a more severe scrutiny. It consequently behoves every dyer or manufacturer, wishing to secure a share of this business, to make himself thoroughly familiar with the subject of khaki-dyeing in all its aspects, and he must indeed have already reached a high standard of excellence if, after a patient and careful perusal of these pages, he has not added something to his knowledge, or has not been inspired with some idea towards further progress.—

June 1903.

Index.

Introduction.

The word „**Khaki**“, sometimes spelled „Khakee“, „Khake“ or „Kaki“, is currently used to describe certain shades of drab, varying from grey to olive, and from olive to brown.

As to the origin and meaning of this word, by general consent it comes from India, and signifies „earth colored“.

Whilst garments of this modest and serviceable hue were no doubt worn extensively in remote ages, the late revival is entirely due to the manner, in which this shade meets the requirements of modern warfare and tactics. The ever increasing range and accuracy of firearms render it desirable, that the combatants should, by adapting themselves to the surroundings, offer the worst possible target to the ennemy, and though the results of experiments, made of late in various countries, for determining the exact color, most suitable in that respect, may not entirely agree, there seems to be no doubt, that such tints, as are covered by the term „khaki“, are all-round well adapted to the purpose. It may accordingly be assumed, that for army-clothing „khaki“ will remain in permanent and growing demand, though there may be a slight departure from the particular tint, so much in evidence during the late China and Transvaal wars.

The true khaki of recent times, meaning, in a strictly technical sense, a coloration produced, by compounding on the fibre certain metallic oxides, though incomparably fast to exposure, is nevertheless by no means perfect, as regards invisibility. An ideal color in that sense is the much darker shade, presented by the plumage of the common sky-lark (alauda arvensis), such as is reproduced in our pattern No. 26 page 77, supplied by the „Farbenfabriken“, Elberfeld. The same firm have kindly illustrated in pattern No. 27 page 78 on cotten-cloth a shade called „fieldgrey“ (Feldgrau), which has found favour with the German army authorities, whilst pattern No. 62 page 129, due to the kindness of Messrs Leopold Cassella & Co. shows the same shade on wool. Another pattern No. 54 page 121, also due to favour of the „Farbenfabriken“, presents on woollen cloth an olive-grey, which is about to replace the former khaki in the American army. As regards fastness to exposure, it may however, be summarily said, that none of these later shades can compare with the original metallic-oxide khaki.

Some of the methods employed for „khaki dyeing“, even in recent times, were exceedingly crude and primitive.

If we read, that at the outbreak of the last war in Afghanistan English troops stained their white uniforms in a bath of cowdung, sudden emergency may he pleaded in extenuation. It is certainly more surpising, that the German navy-administration should, in fitting out the recent expedition to China, have consumed considerable quantities of such a preparation, as that invented by **Milch** of Wilhelmshaven, which consisted of nothing more nor less than a **decoction of roasted chicory and chlorophyll** (vegetable green). This liquid preparation was supplied in tins, and white, tropical garments could readily be stained by dipping and airdrying, with that peculiar advantage, of being again easily stripped on repeated soaping. A decoction of roasted coffee and green leafs was to serve as a substitute in case of need.

The use of chicory as a dye had in any case been anticipated, **an English patent No. 13214** having already been granted in 1885 **to Gaillot-Lake** for an invention described as follows:

— „This invention relates to the dyeing of silk, wool and other substances, and comprises an improved process, which consists essentially in the employment of a bath, composed of chicory root, browned in an oven and pounded; the proportions used are about one glass of the chicory powder to each litre of water. The water is first boiled and then chicory is added thereto, and the whole is boiled for about six minutes, and is then put into the bath. When this operation has been effected, the bath is again brought to the boiling point, and the objects to be dyed are plunged therein and are constantly turned.

The said objects gradually receive the colour. It requires two or three minutes for straw-colour, and the longer the duration of the immersion, the darker will be the colour obtained. A dark chestnut-brown can, if desired, be thus obtained.

The objects, when taken out of the bath, are immediately steeped in a bath of fresh water, and are then again dried; great brilliancy is thus imparted to silks and woollen goods.

Instead of chicory root, packets of chicory, such as are used in commerce, may be employed.

This dyeing has been tested and does not change. It is applicable to silk, wool, cotton, to any textile fabric, to felt, straw and other materials." —

As already stated, the word „khaki“ in a general sense describes a certain shade, or rather a range of shades. Technically hower, and in the original English meaning, it is synonymous with one particular coloring on cotton, distinguished by a degree of fastness, hitherto unattainable by any natural or artificial true dyestuff, and whereto for wearing qualities even the modern sulphurcolors must yield precedence.

This, the true and original „khaki“, which shall be discussed first and foremost, consists of a mixture merely of the oxides of heavy metals, those of chromium and iron in particular.

Fixation of these oxides upon the fibre offers in itself nothing essentially new; nevertheless it has, in England mainly, become the subject-matter of quite a special litterature, patent-litterature in particular.

In the present pages the whole of this material shall be reviewed in the light of personal experience and with the help of dyed patterns, produced in actual practice.

For the convenience of the reader a few tables, comparing the various hydrometer and thermometer scales and showing the specific gravity and concentration of the principal solutions employed, shall be inserted in the first instance.

Table I.

Comparative Hydrometer Scale.

Specific Gravity, degrees Beaumé & Twaddle at 12.5° C.

Sp. Gr.	Beaumé	Twaddle	Sp. Gr.	Beaumé	Twaddle
1·000	0 deg.	0·0 deg.	1·252	29 deg.	50·4 deg.
1·007	1 „	1·4 „	1·263	30 „	52·6 „
1·014	2 „	2·8 „	1·274	31 „	54·8 „
1·022	3 „	4·4 „	1·285	32 „	57·0 „
1·029	4 „	5·8 „	1·297	33 „	59·4 „
1·037	5 „	7·4 „	1·308	34 „	61·6 „
1·045	6 „	9·0 „	1·320	35 „	64 „
1·052	7 „	10·4 „	1·332	36 „	66·4 „
1·060	8 „	12·0 „	1·345	37 „	69 „
1·067	9 „	13·4 „	1·357	38 „	71·4 „
1·075	10 „	15·0 „	1·370	39 „	74 „
1·083	11 „	16·6 „	1·383	40 „	76·6 „
1·091	12 „	18·2 „	1.397	41 „	79·4 „
1·100	13 „	20·0 „	1·410	42 „	82 „
1·108	14 „	21·6 „	1·424	43 „	84·8 „
1·116	15 „	23·2 „	1·438	44 „	87·6 „
1·125	16 „	25 „	1·453	45 „	90·6 „
1·134	17 „	26·8 „	1·468	46 „	93·6 „
1·142	18 „	28·4 „	1·483	47 „	96·6 „
1·152	19 „	30·4 „	1·498	48 „	99·6 „
1·162	20 „	32·4 „	1·514	49 „	102·8 „
1·171	21 „	34·2 „	1·530	50 „	106 „
1·180	22 „	36·0 „	1·546	51 „	109·2 „
1·190	23 „	38·0 „	1·563	52 „	112·6 „
1·200	24 „	40·0 „	1·580	53 „	116 „
1·210	25 „	42·0 „	1·597	24 „	119·4 „
1·220	26 „	44·0 „	1·615	25 „	120·3 „
1·231	27 „	46·2 „	1·634	56 „	126·8 „
1·241	28 „	48·2 „			

Table II.

Comparative Hydrometer Scale.

Degrees Twaddle and Beaumé, and Specific Gravity. at 12,5° C.

Twaddle	Beaumé	Sp. Gr.	Twaddle	Beaumé	Sp. Gr.
0 deg.	0 deg.	1·000	48 deg.	27·9 deg.	1·240
1 „	0·7 „	1·005	49 „	28·4 „	1·245
2 „	1·4 „	1·010	50 „	28·8 „	1·250
3 „	2·1 „	1·015	51 „	29·3 „	1·255
4 „	2·7 „	1·020	52 „	29·7 „	1·260
5 „	3·4 „	1·025	53 „	30·2 „	1·265
6 „	4·1 „	1·030	54 „	30·6 „	1·270
7 „	4·7 „	1·035	55 „	31·1 „	1·275
8 „	5·4 „	1·040	56 „	31·5 „	1·280
9 „	6·0 „	1·045	57 „	32·0 „	1·285
10 „	6·7 „	1·050	58 „	32·4 „	1·290
11 „	7·4 „	1·055	59 „	32·8 „	1·295
12 „	8·0 „	1·060	60 „	33·3 „	1·300
13 „	8·7 „	1·065	61 „	33·7 „	1·305
14 „	9·4 „	1·070	62 „	34·2 „	1·310
15 „	10·0 „	1·075	63 „	34·6 „	1·315
16 „	10·6 „	1·080	64 „	35·0 „	1·320
17 „	11·2 „	1·085	65 „	35·4 „	1·325
18 „	11·9 „	1·090	66 „	35·8 „	1·330
19 „	12·4 „	1·095	67 „	36·2 „	1·335
20 „	13·0 „	1·100	68 „	36·6 „	1·340
21 „	13·6 „	1·105	69 „	37·0 „	1·345
22 „	14·2 „	1·110	70 „	37·4 „	1·350
23 „	14·9 „	1·115	71 „	37·8 „	1·355
24 „	15·4 „	1·120	72 „	38·2 „	1·360
25 „	16 „	1·125	73 „	38·6 „	1·365
26 „	16·5 „	1·130	74 „	39·0 „	1·370
27 „	17·1 „	1·135	75 „	39·4 „	1·375
28 „	17·7 „	1·140	76 „	39·8 „	1·380
29 „	18·3 „	1·145	77 „	40·1 „	1·385
30 „	18·8 „	1·150	78 „	40·5 „	1·390
31 „	19·3 „	1·155	79 „	40·8 „	1·395
32 „	19·8 „	1·160	80 „	41·2 „	1·400

Twaddle	Beaumé	Sp. Gr.	Twaddle	Beaumé	Sp. Gr.
33 deg.	20·3 deg.	1·165	81 deg.	41·6 deg.	1·405
34 „	20·9 „	1·170	82 „	42·0 „	1·410
35 „	21·4 „	1·175	83 „	42·3 „	1·415
36 „	22·0 „	1·180	84 „	42·7 „	1·420
37 „	22·5 „	1·185	85 „	43·1 „	1·425
38 „	23·0 „	1·190	86 „	43·4 „	1·430
39 „	23·5 „	1·195	87 „	43·8 „	1·435
40 „	24·0 „	1·200	88 „	44·1 „	1·440
41 „	24·5 „	1·205	89 „	44·4 „	1·445
42 „	25·0 „	1·210	90 „	44·8 „	1·450
43 „	25·5 „	1·215	91 „	45·1 „	1·455
44 „	26·0 „	1·220	92 „	45·4 „	1·460
45 „	26·4 „	1·225	93 „	45·8 „	1·465
46 „	26·9 „	1·230	94 „	46·1 „	1·470
47 „	27·4 „	1·235	95 „	46·4 „	1·475

Table III.

Comparative Thermometer Scale.
Degrees Réaumur, Fahrenheit und Celsius (Centigrade).

Réaumur	Fahrenheit	Celsius	Réaumur	Fahrenheit	Celsius
+80	+212	+100	+47	+137·75	+58·75
79	209·75	98·75	46	135·50	57·50
78	207·50	97·50	45	133·25	57·25
77	205·25	96·25	44	131	55
76	203	95	43	128·75	53·75
75	200·75	93·75	42	126·50	52·50
74	198·50	92.50	41	124·25	51·25
73	196·25	91·25	40	122	50
72	194	90	39	119·75	48·75
71	191·75	88·75	38	117·50	47·50
70	189·50	87·50	37	115·25	46·25
69	187·25	86·25	36	113	45
68	185	85	35	110·75	43·75
67	182.75	83·75	34	108·50	42·50
66	180·50	82·50	33	106·25	41·25
65	178·25	81·25	32	104	40
64	176	80	31	101·75	38·75
63	173·75	78·75	30	99·50	37·50
62	171·50	77·50	29	97·25	36·25
61	169·25	76·25	28	95	35
60	167	75	27	92·75	33·75
59	164·75	73·75	26	90·50	32·50
58	162·50	72·50	25	88·25	31·25
57	160·25	71.25	24	86	30
56	158	70	23	83·75	28·75
55	155·75	68·75	22	81·50	27·50
54	153·50	67·50	21	79·25	26·25
53	151·25	66·25	20	77	25
52	149	65	19	74·75	23·75
51	146·75	63·75	18	72·50	22·50
50	144·50	62·50	17	70·75	21·25
49	142·25	61·25	16	68	20
48	140	60	15	65·75	18·75

Table IV.

Specific Gravity of Carbonate of Soda solutions at 15° C. (according to Lunge).

Specific Gravity	Degrees Beaumé	Degrees Twaddle	Percentage by weight		1 metre-cube contains Kilos	
			Na_2CO_3	Na_2CO_3, 10 aq	Na_2CO_3	Na_2CO_3 + 10 aq
1·008	1	1·4	0·67	1·807	6·8	18·2
1·014	2	2·8	1·33	3·587	13·5	36·4
1·022	3	4·4	2·09	5·637	21·4	57·6
1·029	4	5·8	2·76	7·444	28·4	76·6
1·036	5	7·2	3.43	9·251	35·5	95·8
1·045	6	9·0	4.29	11·570	44·8	120·9
1·052	7	10·4	4·94	13·323	52·0	140·2
1·060	8	12·0	5·71	16·400	60·5	163·2
1·067	9	13·4	6·37	17·180	68·0	183·3
1·075	10	15·0	7·12	19·203	76·5	206·4
1·083	11	16·6	7·88	21·252	85·3	230·2
1·091	12	18·2	8·62	23·248	94·0	253·6
1·100	13	20·0	9·43	25·432	103·7	279·8
1·108	14	21·6	10·19	27·482	112·9	304·5
1·116	15	23·2	10·95	29·532	122·2	329·6
1·125	16	25·0	11·81	31·851	132·9	358·3
1·134	17	26·8	12·61	34·009	143.0	385·7
1·142	18	28·4	13 16	35·493	150·3	405·3
1·152	19	30·4	14·24	38·405	164·1	442·4

Table V.

Specific Gravity of Caustic Soda at 15° C. (acc. to Lunge).

Specific Gravity	Degrees Beaumé	Degrees Twaddle	Percentage by weight		1 meter cube contains Kilos	
			Na_2O	NaOH	Na_2O	NaOH
1·007	1	1·4	0·47	0·61	4	6
1·014	2	2·8	0·93	1·20	9	12
1·022	3	4·4	1·55	2·00	16	21
1·029	4	5·8	2·10	2·71	22	28
1·036	5	7·2	2·60	3·35	27	35
1·045	6	9·0	3·10	4·00	32	42
1·052	7	10·4	3·60	4·64	38	49
1·060	8	12·0	4·10	5·29	43	56
1·067	9	13·4	4·55	5·87	49	63
1·075	10	15·0	5·08	6·55	55	70
1·083	11	16·6	5·67	7·31	61	79
1·091	12	18·2	6·20	8·00	68	87
1·100	13	20·0	6·73	8·68	74	95
1·108	14	21·6	7·30	9·42	81	104
1·116	15	23·2	7·80	10·06	87	112
1·125	16	25·0	8·50	10·97	96	123
1·134	17	26·8	9·18	11·84	104	134
1·142	18	28·4	9·80	12·64	112	144
1·152	19	30·4	10·50	13·55	121	156
1·162	20	32·4	11·14	14·37	129	167
1·171	21	34·2	11·73	15·13	137	177
1·180	22	36·0	12·33	15·91	146	188
1·190	23	38·0	13·00	16·77	155	200
1·200	24	40·0	13·70	17 67	164	212
1·210	25	42·0	14·40	18·58	174	225
1·220	26	44·0	15·18	19·58	185	239
1·231	27	46·2	15·96	20·59	196	253
1·241	28	48·2	16·76	21·42	208	266
1·252	29	50·4	17·55	22·64	220	283
1·263	30	52·6	18·35	23·67	232	299
1·274	31	54·8	19·23	24·81	245	316
1·285	32	57·0	20·00	25·80	257	332
1·297	33	59·4	20·80	26.83	270	348
1·308	34	61·6	21·55	27·80	282	364
1·320	35	64·0	22·35	28·83	295	381
1·332	36	66·4	23·20	29·93	309	399

Table VI.

Specific Gravity und Concentration of Sulphate of Iron and Sulphate of Copper Solutions at 15° C.

Specific Gravity	Percentage by weight $FeSO^4+7H^2O$	Specific Gravity	Percentage by weight $CuSO^4+5H^2O$
1·011	2	1·0125	2
1·021	4	1·0254	4
1·032	6	1·0384	6
1·043	8	1·0516	8
1·054	10	1·0649	10
1·065	12	1·0785	12
1·082	15	1·0933	14
1·112	20	1·1063	16
1·143	25	1·1208	18
1·174	30	1·1354	20
1·206	35	1·1501	22
1·239	40	1·1659	24

Table VII.

Specific Gravity und Concentration of Chrome-Alum-solutions at 17·5° C. (acc. to Franz).

Percentage by weight $K^2Cr^2(SO^4)^4+24$ aq	Specific Gravity	Percentage by weight $K^2Cr^2(SO^4)^4+24$ aq	Specific Gravity
5	1·0174	40	1·1896
10	1·0342	50	1·2894
20	1·0746	60	1·4566
30	1·1274	70	1·6362

The **first** English patent on Khaki-dyeing **No. 11456 A. D. 1884** was granted to **Frederick Albert Gatty,** the invention being described as follows:

— „I take a solution of chrome alum or any other soluble salt of chromium and mix therewith a solution of sulphate of iron or any other soluble salt of iron, capable of being precipitated with an alkali. After the yarns or fabrics have been boiled in weak soda ash and washed in the ordinary way well understood by dyers, for the purpose of freeing them from all impurities, and the excess of water has been extracted, the yarns or fabrics are impregnated with the mixed solutions of chromium and iron salts.

For an olive shade I take five gallons of a solution of chrome alum at ten degrees of Twaddle's hydrometer or an equivalent quantity of acetate of chromium or chloride of chromium or any other soluble salt of chromium, and mix therewith one gallon of a solution of sulphate of protoxide of iron, containing one and half pounds weight Avoirdupois of the sulphate per gallon, or an equivalent quantity of any other soluble salt of iron, such as the acetate, or common iron liquor.

For brown shades, I increase the quantity of the iron salt according to the shade required, and extract or squeeze out the excess of the liquid, in the case of yarns by wringing, and in the case of fabrics by passing them between two rollers, or in any other suitable manner. The yarns of fabrics are then passed through a solution of soda at four or more degrees of Twaddle's hydrometer; this solution I prefer to use at 140° to 150°

Fahrenheit, but a higher or lower temperature does not materially affect the result.

In place of soda, potash, ammonia, or their carbonates may be used. If ammonia is used, it must be applied cold or nearly so.

After coming out of the soda or other alkaline solution, the yarns or fabrics are washed in water and then dried, but if the colour is found too pale, the process may be repeated a second or third time, until the shade is dark enough. For very full shades, I prefer to dry the yarn and fabrics before impregnating them with the mixed solution of chromium and iron salt, and dry them again, after being impregnated, by running them over dyring cylinders, or in any other suitable way. In this case, in order to prevent the yarns or fabrics from being injured during the drying operation, I prefer to use acetate of chromium or a mixture of acetate of chromium and chrome alum and acetate of iron at ten to fifteen degrees of Twaddle's hydrometer, according to the shade desired. After being dried, the yarns or fabrics are passed through the solution of soda or other alkali, as stated above.

If the dry yarns or fabrics, where the acetate of chromium has been used, are steamed, the passage through soda may be dispensed with, but I prefer to pass them through the alkaline solution.“ —

— It will be the more expedient to communicate at this early stage some personal observations on fixing khaki dyeings, as these were made available in the production of the majority of patterns, contained in this book.

(1). The dyeings produced in the manner indicated by Gatty's patent are not fast to acid.

Half-an-hour steeping in dilute 5 % sulphuric acid (equal to 7.2° Tw.) effects complete decolorisation.

(2). A passage through the Matter-Platt ager causes no improvement in this respect.

(3). Half-an-hour to one hour's steaming at 15 lbs. pressure improves the resistance to acid to a notable, still by no means satisfactory, extent.

(4). If the steaming of goods, padded with acetate of iron and acetate of chrome and dried in the hot-flue or on the air, follows fixing with carbonate of soda, the fibre remains much stronger, than under the reversed order of operations.

Thus, bleached cotton-coth, showing in the dynamometer in the direction of the warp an original tensil strength of 47 Kilos, hereafter bore the following strain:

(a.) When steamed before passage through carbonate of soda 29.9 Kos.

(b.) When steamed after passage through carbonate of soda 40 Kos.

This will explain, why khaki - colored cottoncloth's supplied to the German colonial service by certain establishments, employing the former rotation, were found so deficient in strength.

The use of acetate of iron, as compared with pyrolignite of iron, sometimes described as unsuitable for khaki-dyeing, (see Leipziger Färberzeitung 1900, No. 14), has decided disadvantages, regarding the preservation of tensil strength.

The experiments of Schwarz & Schlumberger, (Bull. de la Soc. Ind. de Mulhouse, vol. I pg. 198 and vol. XIII pg. 422), have shown, that cottonfabrics, impregnated with ferrous salts, are, on subsequent atmospheric oxidation, tendered the more thoroughly, the more rapidly oxidation proceeds, and that pyrolignite of iron is in this sense far more innocuous than acetate.

Apart from this, and besides being much cheaper, pyrolignite, thanks to certain protective empyreumatic constituents, is much less liable, on standing in contact with the air, to separating insoluble basic salts, than acetate, and the same may be said of commercial chrome-alum, as compared with acetate of chrome.

These tarry constituents turn the khaki slightly grey and dull, but not to any objectionable extent.

When working with pure acetate of iron, it is customary to protect the stocksolution against atmospheric action by a floating layer of tar.

Neither steaming, nor the passage through alkali, severally or jointly, suffice to render the coloring produced on the above lines, though it meets the most exacting demands for fastness to washing and light, sufficiently acid-proof. Oxalic acid destroys and strips it with particular ease.

As a remedy for this defect **Gatty & Coy Ltd.** and **Victor Herbert Gatty of Preston,** by the English patent **No. 7041 A. D. 1897,** protected the following after-treatment:

—"In order to greatly increase the fastnest, against the action of acids, of colours produced on vegetable fibres, yarns and fabrics by the oxides or insoluble salts of iron and chromium, fixed separately or together, we treat the fibres, yarns or fabrics, upon which the oxides or insoluble salts or one of them have already been fixed, with a solution of a soluble silicate.

For instance, we take the fibres, yarns or fabrics, upon which the oxides or insoluble salts have already been fixed in any suitable way, and subject them to the action of a solution of a soluble silicate, such as silicate of soda. We find that a solution of two degrees Twaddle gives good results, but the strength is not material. We prefer to use this solution at or near boiling point, in which case we find, that an immersion of fifteen minutes in the solution is sufficient to give good results, but the treatment may be continued for an hour; a simple passage of the goods of one minutes duration through the boiling solution is sufficient to materially increase the fastness of the colours against the action of acids. If the solution be used at a lower temperature, a longer immersion is necessary to produce the same result.

The treatment with the silicate solution may be performed in any way convenient or customary in dyeing-operations, or the goods may be simply passed trough the boiling solution and packed together or steamed whilst saturated with it, but we prefer to keep the goods immersed in the boiling solution for at least fifteen minutes.

After this treatment the goods are simply washed and may then be dried. The colours may then be subjected without injury to the action of acids of a strength, which would have entirely altered or destroyed them before treatment.

It must be understood, that this result is not attained by mixing silicate of soda or other soluble silicate in any proportion with the precipitant used to fix the oxides or insoluble salts, (where they are fixed by means of a precipitant), but that the treatment with the soluble silicate is an additional process, subsequent to the fixation, in any suitable way, of the oxides or insoluble salts (or one of them) of iron and chromium upon the fibres, yarns or fabrics.

As examples of suitable methods of fixing the oxides or insoluble salts the following may be cited:—

By impregnation of the fibres, yarns or fabrics with solution of any soluble salts of iron $\frac{\text{and}}{\text{or}}$ of chromium, basic or normal, followed, with or without a previous drying, by fixation with an alkali, such as soda, potash ammonia or their carbonates, or by ammonia gas, or in any other suitable way, or, in the case of salts of iron and chromium containing volatile acids, such as the acetates or bisulphites, by fixation by means of steaming or expelling the volatile acids in any other suitable way, or by depositing the oxides or insoluble salts on the fibres, yarns or fabrics from alkaline solutions of iron $\frac{\text{and}}{\text{or}}$ of chromium."—

—**Another method,** to achieve the same result, forms the subject matter of the English patent **No. 9 A. D 1899** granted to **Gartside & Co.** and **Wm. Warr** of Manchester. These inventors express themselves as follows:

"In carrying our invention into effect, we impregnate the vegetable fibres, yarns or fabrics with an acid solution of a salt of iron or chromium, or a mixture of the two, then dry the goods, and subsequently fix the oxide or oxides by means of an alkali in the manner well understood by dyers.

After this treatment the colour is fairly fast to alkali, but does not sufficiently resist the action of acids. To overcome this defect, we first wash the goods and then treat them with a boiling solution of boracic acid, ammonium borate or its chemical equivalent. The goods may either be boiled in the solution, or be padded and subsequenthy steamed. We find that padding with ammonium borate and subsequent steaming for, say, twenty minutes gives the best results. The padding solution is composed of, say, two and a half ounces of boracic acid and six ounces of liquid ammonia of twenty five *per cent.* strength to every gallon of water; but we use this solution stronger or weaker, according to the depth of colour desired. The goods are then dried and may be subsequently finished by any of the processes well known to dyers." —

— **Comparative examination of these two methods results entirely in favour of Gatty.**

Apart from the point of cost, the treatment with borate of ammonia yields colors, which in no way satisfy demands for superior fastness.

A **third suggestion** in due to **Fürth**, who in Lehne's "Färber-Zeitung" (June 1900) reports thus: "—In order to impart to mixed dyeings of oxide of iron and chromium considerable resistance even to mineral acids, it suffices to merely add to the current process the operation of padding with turkey-red oil, combination with the oil being effected by finally steaming the dried goods under pressure.

In detail the process runs as follows:

The well-boiled or half-bleached goods are given about 6 ends on the jigger, set for 25 gallons with 2 quarts of persulphate of iron (commercial "nitrate of iron" or „iron liquor") of 90.6 Tw.—25 lbs. of chrome-alum previously dissolved,—and 1 pint of pyrolignite of iron — the latter for shading towards grey. This operation being performed towards evening, the batched goods stand over night; next morning, for the purpose of evenly distributing the moisture, turn the roll and, after standing for half-an-hour, give another 2 turns through, the bath. Immediately after, give 2 ends in hot carbonate of soda, 7 oz. per gallon, wash and dry. Next pad in a turkey-red oil solution, containing 2 lbs. of 30% oil per gallon, dry and steam for 1 hour at 8 lbs. pressure. Finish as usual, but with a reduced quantity of fatty matter.

If the work is to be done in a continuous manner, say in the open-soaper, charge the first compartment with treble the above quantity of metallic salts, the second with soda, and wash in the third.

By varying the proportions, the shade may be modified at will, somewhat in the following direction:

	Reddish yellow	Greenish yellow	
"nitrate of iron" 90.6° Tw.	2 quarts	1 quart	For 25 gallons
chrome-alum	27 lbs.	20 lbs.	
pyrolignite of iron 18.2° Tw.	1 pint	1 pint	

	Intermediate shades			
	I.	II.	III.	For 25 gallons
"nitrate of iron" 90.6° Tw.	1 quart	1 quart	3 gills	
chrome alum	30 lbs.	20 lbs.	25 lbs.	
pyrolignite of iron 18.2° Tw.		2 quarts	1 gill	

Even Fürth's method will however not bear comparison with Gattys silicate process.

Thus, whilst the subjoined pattern No. 1 fol. 24, treated with silicate after steaming, remains practically unchanged in 5 percent sulphuric acid, pattern No. 2 fol. 24, fixed with oil in the manner described, is almost entirely decolorised The same applies to patterns No. 3 and No. 4.

Hereafter, whenever allusion is made to **after-treatment,** it will **exclusively mean fixing with silicate of soda.**

As has already been shown on page 15, the order, in which the operation of steaming takes place, is of the greatest possible consequence. This fact will be rendered still more striking by the following table:

The same khaki treated as:		Steeped for 1/2 hour in 50 cc. H_2SO_4 to 1 quart
1	Neither steamed nor fixed	Completely decolorised
2	Steamed but not fixed	Completely decolorised
3	Fixed with silicate and steamed	Semi-discharged
4	Steamed and fixed with silicate	Almost unchanged

Hence it results that the **passage through silicate should always follow steaming!**

Patterns No. I to No. 4 (incl.), were prepared in order,-firstly, to show the shades produced with various proportions,-secondly, that they might be submitted to a comparative test for solidity to acids-and, last not least, to determine the influence of the oiltreatment upon the feel of the cloth and its fitness for sewing.

It is well known that genuine khaki-dyed cotton-cloth's are exceeding bad to sew; this defect, which shall be more fully dealt with in the sequel, is due to accumulation of metallic oxides upon the fibre.

No. I pattern has been produced as follows:

Dyebath:

chrome alum		$73\frac{1}{4}$ lbs.	
persulphate of iron (comm. nitrate)	90.6° Tw.	3	quarts
pyrolignite of iron	25° Tw.	5	quarts
		to make 25 gallons.	

The bleached fabric was on the padding machine passed 4 times through the above cold bath, then again on another padding machine 4 times through boiling soda, (7 oz. of calcined carbonate of soda per gallon), and hereupon washed in the open-soaper. The **entire** manipulation was **repeated 3 times,** without any intermediate drying.

The cloth was then dried on the stenter, steamed for 1 hour at 15 lbs. pressure, given 6 ends in boiling silicate of soda (7 oz. of 52° Tw. per gallon), slightly soaped and finally dried.

The original shade is slightly reddened, both by the steaming and the treatment with silicate.

The shade thus produced is of course final, as any attempt to modify it by topping would necessarily interfere with its solidity. This is, in fact, the critical point of khakidyeing. In the case of army orders, demanding an exact match, the strictest possible control of the chromium and iron compounds employed and the most scrupulous adherence to the formula only, can ensure success. These conditions observed, there is no further uncertainty, difficulty or risk.

No. 2 pattern has been dyed in exactly the same manner as No. 1 pattern. After drying the cloth was padded in alkaline turkey-red-oil (2 lbs. per gallon), steamed for 1 hour at 15 lbs. pressure, rinsed and slightly soaped.

To arrive at a serviceable shade with both No. 1 and No. 2 patterns **the entire process** had accordingly to be **repeated 3 times.**

According to Fürth the **continuous process** demands something like treble the quantity of iron and chromiumsalts prescribed on pg. 20 for jiggdyeing. On these lines **No. 3 and No. 4 patterns** were produced, the proportions employed being the following:

No. 3 pattern

chrome alum	88 lbs.
persulphate of iron 90.6° Tw. .	6 quarts
pyrolignite of iron 25° Tw. . .	3 pints
	to make 25 gallons

No. 1.

No. 2.

No. 3.

No. 4.

No. 4 pattern

chrome-alum	66 lbs.
persulphate of iron 90.6° Tw. .	3 quarts
pyrolignite of iron 25° Tw. . .	3 pints
	to make 25 gallons

Also these colors were developed on the padding-machine in the manner described, but by **one single series of operations,** and hereafter fixed with oil.

They are obviously much too light, to be of practical use.—

The next patterns, **Nr. 5 to No. 10** incl., are intended to show further shades, produced by varying the composition of the bath, and moreover **the effect of interposed drying.**

In regard to this latter manipulation, I would offer the following observations:

Drying pieces, impregnated with iron- and chromium-salts, on cans, generally produces uneven and streaky colors; airdrying, with stationary suspension from lath's or rods, causes corresponding marks. The only rational method therefore is drying in the hot-flue. Where such a contrivance is not available, the difficulty has to be met by merely repeating the operation of mordanting and fixing (developing). In this way indeed the most level colors can be got, whilst there is no perceptible difference as regards fastness.

No. 5 and No. 6 patterns

Mordant:	chrome-alum	66 lbs.
	persulphate of iron 90.6° Tw.	5 quarts
	pyrolignite of iron 25° Tw.	3 quarts
		to make 25 gallons

Process: No. 5 passed 3 times through the bath without drying,

No. 6 passed once through the bath and airdried. Both shades developed and aftertreated as formerly described.

No. 7 and No. 8 patterns.

Mordant:	chrome-alum	66	lbs.
	persulphate of iron 90.6° Tw.	3	gallons
	pyrolignite of iron 25° Tw.	2	gallons
	to make	25	gallons.

Process: No. 7 dyed and fixed as No. 5.

No. 8 treated as No. 6.

A few intermediate shades were produced by mixing the above mordants.

No. 9 pattern: grounded in a mixture of equal parts of liquors used for No. 5 and No. 7; otherwise treated exactly as No. 5.

No. 10 pattern: grounded as No. 9; otherwise corresponds with No. 6.

— Before proceeding with these experiments, something shall be said here upon **the test method employed for comparing the solidity of various dyeings.**

These tests consist in:

(1). 6 weeks exposure in the open air, the patterns being stretched upon a frame.

(2). Steeping for 1 hour in a cold solution of $2\frac{1}{2}$ drams of sodium-peroxide and $7\frac{1}{2}$ drams of Epsom-salt per 1 quart.

(3). Boiling for 1 hour with $12\frac{1}{2}$ drams of Castil-soap

No. 5.

No. 6.

No. 7.

No. 8.

and 5 drams of calcined carbonate of soda per 1 quart.

(4). Steeping for 1 hour in 3 percent acetic acid at 40^{0} C. This test was inserted as equivalent to resisting negroperspiration.

(5). Steeping for 1 hour in cold dilute hydrochloric acid 1:10.

(6). Steeping for $^{1}|_{2}$ hour at 40^{0} C. in a 1.5 $^{0}|_{0}$ solution of oxalic acid.

No. 9.

(7). Steeping for 1 hour in clear chloride of lime solution 1.4^{0} Tw. acidulated with $^{1}|_{3}$ oz. acetic acid per quart.

With exception of the exposure test, this ordeal may appear brutal. Its purpose however is not, to show the degree of solidity, reasonably to be demanded from mode-shades of the depth of khaki, but to arrive at a forcible distinction between the particular colors under review, which, if submitted to conventional tests for fastness to washing, perspiration, acid and chlorine, would all of them be found equally perfect. Incidentally these tests will also serve to distinguish genuine khakis from those, not produced with metallic oxides.

The 10 shades hitherto discussed and sampled severally behave as follows:

No. 10.

(1). Six weeks exposure in the open air (June 1 to July 15) hardly affects patterns No. 1—5—6—7—8—9—10; No. 2 turns somewhat lighter, No. 3 and No. 4 somewhat darker, towards grey.

(2). Peroxide of hydrogen has no effect on any of these dyeings.

(3). Boiling soap and soda has no effect.

(4). Except on No. 2, which turns somewhat lighter, acetic acid has no action.

(5). Hardly touched by hydrochloric acid: No. 1—5—6—7—8—9—10; strongly reduced No. 2, less so No. 3 and No. 4.

(6). Oxalic acid slightly attacks pattern No. 1; semi-discharges No. 5—6—9 and 10; almost entirely strips No. 2—3 and 4, the two latter leaving a weak reddish grey; No. 7 and 8 behave somewhat better than No. 5 and 6.

No. 11.

(7). Chloride of lime has little effect, beyond turning the shade somewhat redder. Pattern No. 2 however turns considerably lighter, attaining almost the shade of our pattern No. 17 (pg. 48).

The next series of **patterns No. 11 to No. 14 incl.** were produced with the acetates of chromium and iron, the former, as supplied by de Haen, Hannover, the latter got by interaction between acetate of lead and sulphate of iron (green copperas).

No. 11 and No. 12 patterns:

Mordant: acetate of chrome 32.4° Tw. 5 gallons
acetate of iron 20° Tw. . 2.5 „
water . . . . 5 „

Process: No. 11 produced, as shown with No. 5, in 3 operations, without intermediate drying.

No. 12.

No. 12 produced in 1 operation, with air-drying, as No. 6. Aftertreatment as before.

No. 13 and No. 14 patterns.

Mordant: acetate of chrome 32.4° Tw. 5 gallons
acetate of iron 20° Tw. . 2,5 „
sulphate of copper . . . 3,5 lbs.
water . . . . 5 gallons.

Process: For No. 13 as for No. 11.

For No. 14 as for No. 12, including aftertreatment.

These 2 patterns, when compared with No. 11 and 12, show the effect of added coppersulphate.

—The grant of an English patent **No. 3732 A. D. 1886** to **Henry Kershaw** was successfully opposed.

It will appear from the subjoined text of the speci-

No. 13.

fication in how far the process in question infringes Gatty's patent of the year 1884 or otherwise lacks novelty:

—"My invention relates to the dyeing of cotton and other vegetable fibres and of goods and yarn made therefrom. According to my invention, I obtain fast colours by treating the goods with certain chemical salts. In the carrying into effect of my said invention, I dissolve chromium-sulphate and iron peroxide-sulphate in water, and to the solution I add sodium-bi-carbonate or commercial soda, the quantity of alkali used being such, as that precipitation of oxides does not take place. In practice I have found the following named proportions to give good results. In about forty five gallons of water seventy seven pounds of chromium sulphate and eight and a half pounds of iron peroxide sulphate are dissolved. To this solution about thirty six pounds of sodium-bi-carbonate are added, or as much as the solution will take up without precipitation

No. 14.

taking place. The goods, say for example cotton piece goods, are thoroughly saturated with this solution by immersion or by passing the goods through the solution or by other suitable means, and are then squeezed and dried by means of drying cylinders or are otherwise suitably dried. The goods, are then passed into or through, or treated with, a weak boiling solution of silicate of soda or through a weak solution of soap, the solution used being preferably in a boiling state. By these means I obtain a drab colour, the silicate of soda solution producing greener shades than the soap solution. In place of the latter I may use a weak solution of carbonate of soda, but I prefer the soap solution. I do not confine myself to the sulphate salts, as I may use other salts of chromium and iron-peroxide, which by the addition of carbonate or bi-carbonate of soda form soluble basic salts, or I may use other means to produce the soluble basic salts by reaction upon the chromium and iron and peroxide salts. The proportions, herein before mentioned, are given by way of example, and may be varied according to the shade of colour, which the operator desires to produce.“—

Another innovation in the dyeing-process, of doubtful value, is embodied in **Gatty's** English patent **No. 6128 A. D. 1887.** The patentee expresses himself as follows:

—„After the yarns or fabrics have been boiled and freed from impurities, they are impregnated with a solution, containing for every gallon of water from one to one and a half pounds of bi-chromate of soda or an equivalent quantity of bi-chromate of potash, and from four to six ounces of acetate of soda, or its equivalent carbonate of soda or other alkaline salt, for every pound of bi-chromate of soda or bi-chromate of potash used.

After the excess of liquor has been squeezed out of the materials, they are exposed to the action of sulphurous acid gas, until the chromic acid is desoxidised. In the case of yarns, they are hung upon rods or otherwise conveniently suspended in a chamber filled with sulphurous acid gas, obtained by burning sulphur or by any of the known methods. For fabrics, I find, the most convenient way is, to run them over rollers through a chamber, supplied with sulphurous acid gas as above described.

In order to vary the shades from the green to a brown colour, I add to the solution of bi-chromate of soda or bi-chromate of potash a certain quantity of a metallic oxide, such as a salt of iron, by preference acetate of iron or common iron liquor. The proportion of iron salt used obviously depends upon the colour required. For example, when a green shade is desired, I use about one pint of iron liquor at twenty degrees of Twaddle's hydrometer for every pound of bi-chromate of soda or bi-chromate of potash, while for brown shades I use from a quart to three pints.

After the chromic acid has been desoxidised, which can easily be seen by its green colour, the materials are passed through an alkali or alkaline salt, such as carbonate of soda, but any alkali, which will precipitate the metallic oxides, will answer the purpose. I prefer to use caustic soda at four to six degrees of Twaddle's hydrometer at a temperature of 140° to 160° Fahrenheit, but stronger or weaker alkali does not affect the colour.

In the case of fabrics, if the colours be dark and strong, I prefer to dry them, before passing them through the alkali.

The yarns or fabrics, into or upon which the oxide of chromium, either alone or in combination with other metallic oxides, has been fixed, may be dried, and the oxides used either as self-colours or as mordants for a subsequent dyeing process with colouring matter, such as alizarine or dyewoods." —

—Equally untenable as Kersharw's, proved English patent **No. 15433 A. D. 1894,** originally held by **Hugh Wylie and James Morton,** whereby the wellknown dissociability of metallic sulphites on drying or steaming was to be monopolised. The specification reads thus;

—"This invention has for its object, by an improved process, to fix oxide of chromium and oxide of iron in various proportions on cotton yarns and fabrics, in a manner to produce fast and permanent colors, ranging from olive green to olive brown in shades of various depths.

In carrying out the invention, a solution of bisulphite of chromium is mixed with a solution of bisulphite of iron, and the goods to be dyed are passed through the mixture, which is by preference cold. By way of example, it may be stated that, for a greenish olive, the mixture may consist of two gallons of a solution of bisulphite of chromium of a strength of 12 degrees Twaddle, and one gallon of a solution of bisulphite of iron of the same strength. For browner olives, the proportion of bisulphite of iron should be increased.

A woven fabric may be padded in the usual way between rollers; and yarns may be tramped and wrung in the usual way. On subsequently simply drying or steaming, or drying and steaming the goods, the oxides of chromium and iron become fixed on them. When steaming is employed, it may in some cases be advantageous to introduce some ammonia into the steaming box.

If found desirable, or in order to render the fixation complete, the goods may be passed through a hot bath, containing phosphate of soda, and carbonate of soda or of potash, or caustic soda, as is done in various dyeing operations of an analogous kind."—

—In the scramble for patents, other anorganic compounds, yielding colored oxides, were by no means overlooked. Thus, already in **1886 Gatty** by the English patent **No. 13362** had emphasised the fact, that **ammoniacal sugar oflead** solutions readily part with their oxide to the cottonfibre; aftertreatment with chromate produces yellow, and if this operation is preceded by some other suitable dyeing process, various compound shades are attained.

The use of **manganese-salts**, in conjunction with chromium and iron, forms the subject-matter of **Hay** and **Moffat's** English patent **No. 7919 A. D. 1893.** That this invention is a combination merely of old and wellknown processes, is sufficiently evident from the following description:

—"Our invention relates to dyeing on cotton fabrics and yarns certain fast and permanent colours, varying in shade from grey to olive and olive brown, and this we accomplish, by fixing on the fabrics and yarns oxides of manganese and chromium for grey shades, and oxides of manganese and iron for olive and brown shades.

In carrying our invention into effect, we take for grey shades a solution of any soluble salt of chromium, and mix therewith a solution of any soluble salt of manganese, capable of being precipitated with an alkaline hypochlorite; for olive and brown shades, we take a solution of any soluble salt of manganese and mix therewith a solution of any soluble salt of iron, capable of being precipitated with an alkaline hypochlorite.

After the fabrics or yarns are bleached in the manner well understood by dyers, the fabrics or yarns are ready for being impregnated with the solutions of chromium and manganese or manganese and iron.

For grey shade, we take a solution of chrome-alum at twelve degrees of Twaddle's hydrometer, or an equivalent of any other soluble salt of chromium, and mix therewith acetate of manganese at five degrees of Twaddle's hydrometer, or an equivalent of any other soluble salt of manganese, in the proportion of ten gallons of chrome-alum to four gallons of acetate of manganese.

For olive and brown shades, we take a solution of acetate of manganese at ten degrees of Twaddle's hydrometer, or an equivalent of any other soluble salt of manganese, and mix therewith acetate of iron at ten degrees of Twaddle's hydrometer, in the proportion of ten gallons of acetate of manganese to one gallon of acetate of iron, but the proportions may be varied according to the shade required. The fabrics or yarns, after being impregnated with the metallic solutions, are hydro-extracted or squeezed between rollers, and are then passed through a solution of alkaline hypochlorite at two degrees of Twaddle's hydrometer, at a temperature of one hundred and fifty degrees Fahrenheit. We prefer to use hypochlorite of sodium, but any other alkaline hypochlorite will answer the purpose.

After coming out of the hypochlorite bath, if the colours are not full enough, the operations are repeated till the desired shades are obtained." —

Further, by Engl, patent **No. 7888 A. D. 1899 Horrocks** claimed the exclusive right to use **a mixture of copper and iron salts** in the following sense:

—"The manner in which I produce this shade or shades, is by fixing or depositing on the fibre of the material to be dyed, mixtures of oxide of copper and of iron, and according to the shades, which I require, I vary the strength and proportions of these two ingredients. The way, in which I deposit these mixtures on the fibre and afterwards fix them thereon, is as follows: —I take a solution of water and copper sulphate, commercially known as "bluestone," say at 8 degrees, or other soluble salt of copper, which will answer the same purpose, and I mix therewith in about equal proportions (more or less) a solution of water and sulphate of iron or other soluble salt of iron at, say, from 2 to 6 degrees, which will precipitate on the addition of an alkali.

The strength and proportions of these two solutions I vary according to the shade of colour I require. I balance one against the other, that is, add more of one or the other, as required to produce the result desired. I then run the fabrics or materials to be dyed through this mixed solution of copper and of iron or steep them in it, and pass them out through a pair of bowls or rollers to squeeze out the excess of liquor, which must be done evenly; or I treat the fabric or yarns in a centrifugal hydro-extractor to remove the excess of liquor. I then pass them, after the last named operation, through an alkali, such as caustic soda, common soda or any of their carbonates or silicates in solution with water at from 2 to

4 degrees, or through a weak solution of ammonia and water at 1 to 3 degrees; the degrees herein refered to being those of Twaddles hydrometer. The goods are then washed and dried. If the shade is not dark enough at one dip or passage through the mixed solutions, I then repeat the above process until the correct shade is arrived at.“—

—**Shades produced on these lines** are represented

No. 15.

by **patterns No. 15 and No. 16,** the following particulare having been observed:

Mordant:

No. 15 pattern:

sulphate of copper . . . 15,5 lbs.
sulphate of iron (copperas) 2,25 „
water . . . . 25 galls.

No. 16 pattern:

sulphate of copper . . . . 15,5 lbs.

sulphate of iron . . . . . . 18 oz.
water . . . . 25 galls.

Process: Both patterns, without intermediate drying, padded 3times in the mordant, passed through boiling soda, washed, dried and steamed for 1 hour at 15 lbs. pressure, fixed with boiling silicate, soaped and dried.

The treatment with silicate, (as in the case of patterns No. 13 and No. 14) does **not** improve the resistance to acids;

No. 16.

hydrochloric acid leaves a pale reddish, oxalic a pale blueish tint merely.

—The "khaki's" hitherto described are without exception produced by using a liquor, where various metallic salts are in mixture, iron being a universal constituent.

Now, whilst there is no difficulty in charging the fibre with one color-constituent, the reddish oxide of iron, there arises on the other hand the problem, **how to introduce**

in the most economical manner the necessary quantity of complementary green chromium oxide?

Under ordinary conditions, that is when using mixed liquors, and particularly when hot-air-drying has to be dispensed with, a considerable portion of chromium is washed away and wasted during the process.

Though it was presumably this consideration, that induced Gatty to speak in his patent No. 7041 A. D. 1897 of a joint or **separate** fixation of iron and chromium oxide, (see page 17) the subject has not otherwise engaged the serious attention of English inventors.

Several German patents, on the contrary, bear upon the deposition upon the fibre of larger quantities of oxide of chromium, and though they deal with the operation as one, preparatory to dyeing with adjective organic coloring matters only, they contain hints, equally serviceable in the production of mineral khaki.

Of particular interest are the efforts of **Rolffs & Co.**, Siegfeld, **(Germ. patent No. 97686,** dated **June I. 1898),** to secure a heavy and level compound-mordant by **successive fixation of chromium and iron oxide,** a result but imperfectly attained by the prior process of Sansone's, where goods, first padded in an alkaline chromiumsolution, are immediately after passed through salts of iron or copper.

The essential passages of Rolff's patent read something like this:

"The bleached cottongoods are on the padding-machine passed through an alkaline solution of oxide of chromium, containing about 1½ oz. of Cr_2O_3 per gallon; they are then permitted to stand for 2 hours and hereupon thoroughly rinsed and dried. Th chromium-mordant thus fixed, the goods an padded in a solution of ferreous or ferric salts, containing per gallon an equivalent of 1½ oz. of $Fe_2 O_3$, fixation of the metal being accomplished by a passage through alkali, either immediate or subsequent to hanging and rapid steaming.

When goods, impregnated with alkaline chromic oxide, are, according to Sansone, passed at once through a bath of sulphate of iron, precipitation of ferreous hydrate upon the fibresurface takes place in the first instance, the gelatinous coating thus formed preventing further penetration of the ironsalt and thus hindering complete decomposition and fixation of the chromiumprepare. This phenomenon is more or less accentuated by the structure of the thread and fabric, and also varies topically in consequence of the mechanical inequalities inseparable from the padding process, with the result, that uniformity is throughout lacking.

It is obvious, that the present process entirely eliminates these uncertainties; it yields in fact a perfectly homogenous and thoroughly fixed compound iron and chromium mordant."—

Both Sansone and Rolffs however had been forestalled by **Schmid**, who already **in 1894** had pointed out **an economical and effective method** for attaining the desired result, namely, passing goods, first prepared in the known manner in Koechlin's alkaline chromesolution and thoroughly washed, through a boiling bath of sulphate of iron (green copperas). This simple process yields, from a bath, containing at most 2 oz. of copperas in 3 gallons, a khaki-shade, closely resembling that of our No. 7 pattern.

In explanation Schmid assumes that Koechlin's mordant yields to the fibre not chromiumhydrate merely, but also, in connection therewith, a certain amount of fixed alkali, not removable by washing; the latter hereafter effects precipitation of a second metallic oxide. In the case of iron-salts, subsequent atmospheric oxidation soon converts the original ferreous hydrate into sesqui-hydrate and thus completes fixation. Incidentally it may here be said, that Schmid, besides priority over Sansone and Rolffs, might also claim the distinction of being the first, who handled a true „oxide khaki“, at a time, when a demand for such a pigment as a self-color was far from being anticipated.

—The **great difference in depth,** which results from using a **mixed bath** as against **dividing the process,** will in a manner be demonstrated by the following experiments, carried out on a small scale.

A.

Pad in

acetate of chrome	32.4° Tw.	100 cc
pyrolignite of iron	25° Tw.	30 cc
water	. . . .	650 cc

Fix in boiling soda.

Shade produced: Slightly lighter than our No. 3 pattern (pg. 25).

B.

Pad in

acetate of chrome	32.4° Tw.	100 cc

caustic soda . 66.4° Tw. 100 cc
water . . . . 50 cc

Let stand for 12 hours, wash, squeeze and pad in

pyrolignite of iron 25° Tw. 30 cc
water . . . . 500 cc

Fix in boiling soda.

Shade produced: Same depth as No. 11 pattern (pg. 32). Incidentally it may be said, that drying, before passing through the iron salt, has, with experiment B, very little effect on the intensity.

The subjoined **patterns No. 17—18—19** were produced in the following manner:

No. 17.

1st mordant: acetate of chrome 23.2° Tw. 10 lbs.
caustic soda . .66.4° Tw. 10 lbs.
water. . . . . 1/2 lbs.

2nd „ 3 1/2 oz. of green copperas in 1 gallon of water.

No. 18.

1st mordant: as for No. 17.

2nd „ 1 gill of pyrolignite of iron 25° Tw. in 1 gallon of water.

No. 19.

1st mordant: as for No. 17.

2nd „ 1 gill of pyrolignite of iron 25° Tw.
7 oz. sulphate of copper in 1 gallon of water.

Process for No. 17—18—19:

Pad the bleached goods in the 1st mordant, wind on

No. 17.

No. 18.

and let stand over night; then wash thoroughly at width and pad in the 2nd mordant. Hereafter take through carbonate of soda, steam, fix in silicate of soda and soap, as described for No. 1 pattern.

These 3 shades enjoy in common a remarkable degree of fastness, excepting to acids. Whilst No. 17 and 18 satisfactorily pass the hydrochloric acid test, both are strongly

No. 19.

attacked by oxalic acid. No. 19 is completely stripped by either acid.

The **introduction of copper** accordingly, and as has moreover been already demonstrated with patterns No. 13, 14, 15 and 16, **undermines the resistance to acids.**

Other solid shades may be produced on the proceeding lines by **topping a chromium oxide bottom** with:

(1). nitrate of nickel, $3^{1}|_{2}$ oz. per gallon; yields a greenish grey,

(2). sulphate of copper $3^1|_2$ oz. per gallon; yields a yellowish green,

(3). mixtures of copper-iron-and nickelsalts; produce greenish grey's.

None of these methods however yield shades of the depth of "fieldgrey", shown on No. 27 pattern.

Another chromemordant, available in khaki dyeing, is that invented by **Gallois (Germ. patent No. 45998 A. D. 1889)** and supplied by the "Farbwerke" Höchst as **"Mordant G A I"** and **„G A II".**—According to Henri Schmid (Bull. de la Soc. de Mulhouse 1899 pg. 257), this mordant, consisting of chromiumchromate and corresponding in the normal state with the formula Cr_2O_3 3 CrO_3, is taken up by the fibre as readily as Koechlins alkaline preparation. The fact, that it does not, like the latter, attack the hands of the operator and avoids shrinkage, has facililated its introduction, particularly with cotton hank dyers.

Hanks are first boiled with soda, then steeped over night in "mordant G A II", (being murio-chromate of chromium), wrung and immediately passed through about 2 oz. of soda per gallon at 50° C. and washed ready for dyeing. Goods thus prepared should, in order to avoid tendering, not be exposed to direct sunlight.

Piecegoods are padded in a mixture of „mordant G A I" (being aceto-chromate of chromium) with glycerine or some other reducing agent, whereupon they are dried, passed through the Mather-Platt ager and washed off in weak carbonate of soda. The latter operation serves the purpose merely of wetting out and is not intended to assist fixation, which, on the contrary, is completed during the short steaming.—

By the **German patent No. 58171 A. D. 1891** Messrs **Kalle & Co.** in Biebrich protected a process, whereby basic **antimoniates or tungstates of metals** are precipitated upon the fabric, by impregnating the same successively with metallic salts and alkaline antimoniates or tungstates. The specification says under others: "Cottonfabrics are treated, for example, thus, that they are padded in chromium chloride of 32° Tw. hung up to dry and passed on the jigger through 1,25 lb tungstate of soda per gallon at 35—40° C., to be subsequently washed and dried. Other fibrous materials are treated in the same way, whilst the salts of alumina, iron, tin, copper, lead, manganese, nickel, cobaltum, uranium and zinc may replace those of chromium."

Another german patent **No. 101481 A. D. 1899**, relating to mordanting vegetable fibres with chromium, and which is sufficiently instructive to deserve a somewhat verbation translation, was granted to **Haber.** The specification says:

"For mordanting cotton with chromium, preparatory to dyeing, both chromates and chromiumoxide-compounds find application. The latter are either in the form of salts, whereof oxide of chromium forms the base, or in the form of chrome-sodium-oxide, when chromium hydrate assumes an acid character.

Chromiumsalts may be used in two ways. Either the goods are impregnated with an adequate amount of the salt, precipitation of the hydrate upon the fibre being brough about in a second bath, containing carbonate of soda, phosphate of soda, lime or the like substances, or the fibre is in the first instance charged with an acid mordant, like tannic acid or soluble oil, and then treated with the chromiumsalt. In the latter case, this acid mordant withdraws from the solution, and holds in combination, part of the metallic oxide, whilst an aftertreatment with lime, soda or a similar precipitant completes the process.

Neither of these methods is entirly satisfactory, for this reason, that the cottonfibre in itself does not possess sufficient affinity to chromium, to with-draw from the solution presented to it an adequate quantity, a state of things, but indifferently remedied by a preparatory treatment with tannin matters or sulphated oil. This defect becomes all the more pronounced, when using dilute solutions of chromiumsalts. An aftertreatment with precipitants is of advantage only in so far, as it tends to convert basic salts already fixed upon the fibre into hydrate. Unmodified chromiumsolution, mechanically adhering, is certainly decomposed, but without fixation of the separated hydrate. The latter adheres but superficially, and on dyeing yields a lake, readily removed by washing and soaping.

The present process obviates these difficulties thus, that the precipitant is incorporated to the fibre in the first instance. When such fibre is subsequently treated with a chromiumsalt, the acid of this salt combines with the precipitant, whilst chromiumhydrate is firmly deposited upon the fibre. The precipitants in question are alkaline earth's, which, in spite of its feeble acid character, are readily fixed by the cottonfibre. For example, 100 kos. of cotton are worked in limewater, squeezed, rinsed and hereafter boiled in a solution of 12 kos. of chromium-fluoride. Prolonged exposure to the air, in order to convert the hydrate of lime into carbonate, may preceed boiling with the chromiumsalt.

If in this process the cotton has been previously prepared with tannic acid or some tanniferous substance, it will be enabled to attract a larger quantity of alkaline earth in form of the tannate, with the result, that subsequent boiling with a chromiumsalt solution, however dilute, will cause a copious fixation of chromiumhydrate. Thus 40 grammes of cottonhank, prepared with 10 percent by weight of tannic acid or its equivalent, steeped in lime-water and rinsed, quantitatively exhaust a solution of 2.4 grammes of chromefluoride in 750 cc of water. On a practical scale, 100 kilos of cotton are mordanted with 10 kos. of tannic acid or with a corresponding weight of sumach or some similar material, passed through lime-water, rinsed and boiled for 2 hours in a solution of 12 kos. of fluoride of chrome.

The present process consequently attains an almost ideal utilisation of the chromiumbath, whilst the liberal and perfect manner, in which the hydrate becomes fixed upon the fibre, leads hereafter to dyeings of unusual depth, brightness and fastness to soap."—

A still more recent **German patent No. 112611 A. D. 1898**

of the **"B. A. & S. F."** describes a new method for the use of **bisulphite of chromium** substantially in the following terms:

"It has been observed that so-called bisulphite of chromium, such as results by dissolving freshly precipitated hydrate in aqueous sulphurous acid, acidulated with sulphuric acid, forms, in conjunction with an aftertreatment with alkaline salts, an exellent cotton mordant.

Hitherto this compound has been employed in such manner, that the fibre impregnated therewith was submitted to drying or steaming, for the purpose of expelling the sulphurous acid and forming basic salts, which on after-treatment with soda or soap yield hydrate of chromium. This process is costly and complicated, and the liberated sulphurous acid, and still more so the sulphuric acid formed by oxidation, are liable to corrode the fibre. Moreover, level results can be got on fabrics only, and the material thus prepared is always difficult to wet out.

The new process, which avoids these difficulties, is based upon the fact, that comparatively dilute solution sof chromium-bisulphite yield to the fibre without drying oxide in a copious and regular manner. Hence the process is particularly adapted for hanks and cottonwool, and it has in the latter case the advantage, of leaving the material open and in good condition for spinning. No trace of sulphurous or sulphuric acid is evolved, hence no risk of tendering. The entire manipulation takes a few hours only, and the material, thus treated, wets out easily.

As against other methods for mordanting cotton with chromiumsalts, our process shows distinct advantages, both in regard to cheapness, simplicity and convenience and in the final result. Nor was this success a foregone conclusion, for whilst many other chromiumsalts, on reaction with soda or the like alkaline salt, also yield hydrate, this is deposited upon the fibre either in a far from regular and even manner, or it adheres too loosely, to withstand rinsing.

On a practical scale we proceed somewhat in the following manner: — 100 kos. of well-boiled and hydroextracted yarn are steeped for some hours, or by preference over night, in chromiumbisulphite of 10° Bé Next wring well, "whiz" and work for 10 minutes at 60° C. in 3 grammes of calcined carbonate of soda per litre. Rinse, again "whiz", oil with a 10 percent solution of turkeyred-oil and dry. The chromiumbath may be kept on and fed with a strong solution, as required.

Whilst the above mordant is calculated for dark colors, a bath of 5° Bé. suffices for medium, one of 3° Bé. for light shades. The strenght of the alkaline fixing-bath may be varied in proportion and, as a matter of course, potassiumcarbonate, sodiumbicarbonate or sodiumphosphate may take the place of soda.

The process, as described, is equally suitable for cottonwool or cottonfabrics and may also be applied to linen, jute, hemp and other vegetable fibres." —

Apart from those relating to the production of pure mineral-oxide-deings, several english patents were also taken, for bringing about otherwise well-known **combinations of tannic acid** or **organic dyestuffs with metallic oxides.**

The first patent of that class, annulled later on, is **No. 11394 A. D. 1887** granted to **Gartside** and **Barnes,** the invention being described as follows:

—"Our invention relates to, and consists in, dyeing cotton, cotton yarn and cotton textile fabrics, and other vegetable fibres and yarns and fabrics composed thereof, with the shades of drab, fawn and other similar colours or shades known as "Khakee" so that such shades or colours shall be "fast".

We take the material, to be operated upon, and immerse it for a suitable length of time in a solution of any suitable kind of tannin matter, such as sumac extract, which solution is made of a strength of about 2° Twaddle. Such solution is made to correspond in strength with the depth or degree of shade or colour required. After immersing the material in the solution of tannin matter aforesaid, we remove the material from the said solution and exhaust or remove the superfluous portion of the aforesaid solution of tannin matter from the material, and afterwards subject the material, to be operated upon, to the action of a dilute solution of iron and chromium. In order to form such solution of iron and chromium, we make a solution of nitrate of iron at or about 2° Twaddle and make a solution of acetate of chromium at or about 3° Twaddle. We then mix the solutions of nitrate of iron and acetate of chromium.—We next wash in water the material, to be operated upon, and subject the material to the action of a boiling solution of bichromate of potash or of bichromate of soda, using a quantity of bichromate of potash or bichromate of soda equal in weigth to about three to four per cent of the weigth of the material. We find in practice that, by adding sulphuric acid to the bichromate of potash or bichromate of soda solution, in a quantity sufficient to liberate nearly the whole of the chromic acid contained in such bichromate of potash or bichromate of soda solution, the action of the chromic acid in such solution is rendered much more rapid. We immerse the material in either of the solutions, until the colour is unalterable and again wash in water the material to be operated upon We next boil the material in a solution of carbonate of soda, such solution being made by employing a quantity of carbonate of soda equal in weigth to about five per cent of the weigth of the material. We boil the material in such solution of carbonate of soda, until the shade of colour of the material undergoes no change. The material is again washed in water and dried, and we afterwards subject the material to the action of steam." —

— On the same level is the English patent **Mitchell-Duxbury, No. 7498 A. D. 1895,** which, besides vegetable fibres, is intended to cover wool. The patentee says:

— "My invention relates to an improved "kaki"-dye for dyeing cotton or woollen goods. The dye is obtained by combining solutions of certain mineral and vegetable matters in stated proportions. The mineral solutions are obtained from the sulphates of copper, zinc, or iron, one or the other being used, according to the shade of "kaki" required.

The vegetable solution or infusion is obtained from myrabolams, divi-divi, gallnuts, catechu or cutch and gambier, redwood, or tanning-bark. These are all well known and regularly imported products, already used largely in tanning and dyeing operations.

In making a kaki dye of an average shade, I would take, say, 2 lbs. of myrabolans or divi-divi, with two ozs. of sulphate of zinc or copper. To these I add one, two or more ozs. of tanning-bark, catechu and redwood and from 10 to 30 or more grains of sulphate of iron, according to the required shade of dye. The foregoing ingredients are mixed with 6 gallons of water, first brought to boiling point, the materials being stirred and soaked for about two hours. Thereafter I add about six ounces of bichromate of potash in six gallons of boiling water, which renders the dye fast.

Variations in the shade of dye can be obtained by varying the proportions of the ingredients. The metallic sulphates being the chief agents in producing the required shade of dye, it will be evident that by varying the proportions of these sulphates, or by u-ing only those sulphates, which give a lighter or darker shade, the dye may be varied. For example, I might use copper-sulphate alone for the lighter shades and iron-sulphate for the darker shades.“ —

As for the intrinsic merits of these "inventions", they merely recapitulate methods, described in the textbooks on dyeing before the middle of the last century; in fact, prior to the introduction of anilines, these methods were the only ones, by which anything like fast shades of this kind could be produced.

These dyeings may be called fast to a certain extent and they were currently so produced on moleskins. A shade got with sumach, yellow catechu, sulphate of iron, chromium and copper, resembling our pattern No. 19, after 9 month's exposure in the open air almost matched pattern No. 22, being however slightly darker.

For a further comparison of these compound shades with true mineral khaki, **pattern No. 20** was produced on the following practical scale:

1st operation: Pad 6 times in a warm solution of $2^1|_4$ lbs. of yellow catechu and 18 oz. of sumach-extract 52.6 degr. Twaddle: in 15 gallons.

2nd operation: Pad 4 times cold or give 4 ends in the jigger in $2^1|_4$ lbs. of green copperas in 15 gallons.

3rd operation: Wash at width.

4th operation: 4 ends on the jigger at the boil through bichromate of potash $1^1|_3$ lbs. and sulphate of copper $1^1|_3$ lbs. in 15 gallons.

5th operation: Rinse, soap warm in 4 oz. of soap per gallon, rinse and dry.

A formula published in the "Dyer and Calico Printer" is this: Boil together 5 lbs. of catechu and 3 lbs. of fustic with sufficient water to make a solution of $3^1|_2$° Bé. Work for 1 hour at 70° C., add $^1|_2$ lb. of bichromate and 2 oz. of green copperas and work for another half-hour.

According to the "Textile Mercury" the fabric is to

steeped for 6—8 hours in a solution of catechu of $1\frac{1}{2}^\circ$ Bé, then passed at 65° C. through a solution of 4 lbs. of bichromate, 1 lb. of green copperas and $\frac{1}{2}$ lb. of sulphate of copper of a concentration of 4° Bé. Hereafter pass first through bisulphite of soda, then through lime-water, rinse and dry.

During the high-tide of khaki dyeing numerous similar

No. 20.

recipes were published, the reproduction whereof would serve no purpose.

None of the colors produced on these lines attain the fastness of true chrome-iron-khaki, nor do they offer any advantage in price, even when the various so-called catechu-substitutes, such as the "Katechin" Reyscher-Barmen, are used in the place of catechu, cutch or gambier.—

Of some slight general interest however might be the

English patent **No. 25581 A. D. 1896** granted to **Eliza Jessie Stewart,** in so far as it draws attention to the tinctorial value of a material not hitherto utilised, namely the shells or husks of the pods of the cotton plant. The invention is described in the following manner:

"My invention relates to improvements in dyeing and which improvements are specially useful in the production of certain fast shades, known as khaki, and in dyeing fabrics in fast colours in browns, drabs and greenish yellows.

I have discovered that, if the outer shells or husks of the pods of the cotton plant, which have hitherto been regarded as a waste-product, be extracted by boiling in water, on decanting and straining the resulting solution, a clear brown liquid is obtained, which on evaporation at a slow rate leaves a glossy brown residue. This clear liquid, or a solution of the residue in water, forms a most valuable dyestuff, not unlike catechu in its properties, for which it may in many cases be used as a substitute, and which is easily oxidised and rendered insoluble by a bath of bichromate of potash or by mere exposure to the air. A convenient strength of extract is obtained by boiling 1 lb. of pods in 16 gallons of water and carefully straining the resulting brown liquid; an extract of about this strength will give good results, though the final colour will be the deeper, the stronger this extract is made. The colouring matter can be obtained in a pure state by evaporating a strong extract of the pods to dryness, and in this condition is more portable than the pods are, and if used in this form, it must be redissolved in the requisite quantity of hot water.

To cause the colour to penetrate the fibre of the

material to be dyed, I may use a bath of myrobalans as a mordant, and to vary the colour produced, I may add salts of iron or copper or other salts to the bath, as is well understood in the working of astringent dyes.

In order to produce the very fast and permanent dye of the peculiar shade. known as khaki, I proceed as follows:—To each gallon of the hot or boiling solution of the extract, obtained from cotton pods as above described, I add about 4 drachms of myrobalans and about $1\frac{1}{2}$ grains of sulphate of iron or other like salt of iron, (the particular quantity of the salt depending on the shade of khaki required, but the more iron is used the greener will be the final shade), and in the bath, either hot or boiling, I steep the fabric to be dyed and then leave it to soak for some hours, after which I pass it through, or rinse it in, water and then either dry it or not. The fabric is then steeped in a warm bath of a weak solution of bichromate of potash, after which it is thoroughly washed, dried and finished. [I have also discovered that in the above process the cotton pod extract may be replaced by catechu or by extract prepared in like manner from the bark of the Babool Tree, and if the same process is carried out, a very fast khaki dye will be produced, and if catechu is employed, a bath of 20 grains of catechu to the gallon will give good results.]

The cotton pod extract is of great value as a colouring matter, in that it is produced from a waste product and is therefore very cheap and easy to abtain. Its shades are easily modified from browns to yellows and yellowish greens by the use of suitable salts; it produces a faster dye than even catechu, and fabrics, steeped in a weak hot solution of the extract and then left to oxidise in the air without further treatment, will be found to have taken a colour, varying from pinkish brown to brown, which is unaltered by washing with soap and even boiling.

As a further instance of the way, in which the cotton pod extract may be used instead of catechu, it will be found that a bath containing 1 oz. of pods and 10 grains of acetate of copper to the gallon, used hot and with a subsequent bichromate of potash bath, will dye a fast brown shade, or with nitrate or sulphate of copper (instead of acetate) a yellowish brown shade will be obtained. This of course is merely the substitution of the extract for catechu in a well known catechu process."—

A **combination of iron- and chromiumoxide** with **alizarine-blue** form the subject-matter of the english patent **No. 3604 A. D. 1898** granted to **John Thomas Reid** and **Henry Thorp,** the manner, in which this combination is effected, being described thus:

—"In carrying our invention into effect, we first scour the fibrous materials, to free them from all impurities, then dry, and when dry may or may not run them through a bath of oleine oil in the proportion of, say, two quarts of oil to sixteen gallons of water, although in practice we have not found this essential, as equally good results are obtained either with or without the oleine oil bath.

We then dry the fibrous materials and afterwards impregnate them with the following dye liquor:—

Twenty gallons of a solution of acetate of chromium (20 degrees Twaddle).

Twenty gallons of a solution of acetate of iron (12 degrees Twaddle).

Eight ounces of alizarine blue S., dissolved in four gallons of a solution of bisulphite of soda or potash (54 degrees Twaddle).

When thoroughly impregnated with the above liquor, we mangle the fibrous materials and then dry them.

We then age the fibrous materials or steam them, and subsequently develop and fix the colour, by passing the fibrous materials through an alkaline bath composed say of

sixteen gallons of water,

eight pounds of common soda,

three gallons of lime water,

but we may employ any other suitable alkali or alkaline solution, except a soluble silicate in the process described in the specification of Gatty's patent numbered 7041 of 1897.

We then by preference soap the fibrous materials and finally wash and dry them.

For full or dark shades we may repeat the above process as often as required, and may vary the proportions for different shades as required.

In place of the alizarine blue S., we might employ ordinary alizarine blue, gallein, coerulein, or like products of anthracene; and in place of the acetates we might employ any other suitable salts of chromium and iron.

Having now particularly described and ascertained the nature of our said invention and in what manner the same is to be performed, we declare, that we are aware of the patent numbered 11456 of 1884 granted to F. A. Gatty, as well as the afore-said patent numbered 7041 of 1897 granted to F. A. Gatty and Company Limited and V. H. Gatty, and we desire it to be understood, that we do not herein make any claim to any of the inventions claimed in the specifications of the said letters patent, but what we claim is:—

"The process of dyeing vegetable fibrous materials certain fast and permanent colours known as "khaki" shades, which process consists essentially in impregnating the fibrous materials with a mixed solution of alizarine blue S., chromium and iron salts, then drying, ageing or steaming, and developing the colour by treatment with an alkali, substantially as herein set forth."—

Sewing true khaki-colored cotton-cloth.

The general adoption of true khaki-colored cotton-cloth is seriously restricted by the difficulty experienced in sewing. The dyeing-process weights the cloth to the extent of about 10 percent, this gain being entirely due to metallic oxides, precipitated in and upon the fibre. These oxides offer to the needle a considerable amount of resistance, to that degree, that they heat or even break it. The seams easily tear, so that khaki-colored coats and other garments will not stand much wear and tear. Even the best English goods show needleholes, where the fibres appear rent, broken or torn, and this holds good also with the best longstaple cotton, so that the material appears by no means suitable for army-clothing. With tents, were the seams are comparatively few and coarse, the objection is not so serious, and it is more than counterbalanced by the extraordinary fastness to atmospheric influence.

It appeared indicated to study to what extent the brittle nature of the fabric might be remedied by the application, before or after dyeing, of certain "softenings". In order to elucidate at the same time the effect of oiling on the shade, parallel-dyeings were made upon bleached cloth, oiled with a 10 percent turkey-red oil solution and airdried **(pattern No. 21)** and not oiled **(pattern No. 22).**

No. 21.

No. 22.

The mordant employed consists of: water 12 gallons, chrome-alum 37 1/2 lbs. nitrate of iron 90.6° Tw. 4 1/2 pints acetate of iron 25° Tw. 3 gills. The goods were passed through carbonate of soda and steamed, as described formerly, but without aftertreatment with silicate, and finally soaped at 50° C. with 4 oz. of soap per gallon.

A comparison shows that **preparatory oiling adds nothing to the depth of the shade, nor could there,** in spite of the soft and oily feel, **be registered any improvement as regards sewing.**

Equally futile in the latter respect proved treatment with a 10 percent glycerin solution after dyeing, or a corresponding addition of glycerine to the mordant, nor does a final hot soaping make any difference.

During the debut of khaki, in Germany at least, considerable confusion prevailed. The **absence of a sound official test** favoured sophistication to such an extent, that even garment-dyers were enabled to share in the army-contracts, whilst the prices tendered were in many instances so low, as to preclude from the outset the use of so laborious and costly a process, as that indispensable for the genuine article. A test actually in use at the time consisted in boiling the fabric in a 4 percent solution of oxalic acid!—Another test was the **spottest with strong hydrochloric or strong nitric acid.** The effect of this test on the dyeings already sampled and on some to follow, shall here be tabulated.

Spottest with strong hydrochloric acid.

No. of pattern operated upon	Reaction	No. of pattern operated upon	Reaction
No. 1	greenish white	No. 20	reddish
2	"	21	greenish white
3	"	22	"
4	"	23	almost unchanged
5	"	24	"
6	"	25	pale red
7	"	26	"
8	"	27	"
9	"	28	"
10	"	29	lighter, yellow
11	"	30	pale red
12	"	31	almost unchanged
13	"	32	"
14	"	33	"
15	"	34	pale red
16	"	35	almost unchanged
17	"	36	pale yellow
18	"	37	pale red
19	"		

Spottest with strong nitric acid.

No. of pattern operated upon	Reaction	No. of pattern operated upon	Reaction
No. 1	light grey, almost white	No. 20	reddish
2	„	21	light grey, almost white
3	„	22	„
4	„	23	almost unchanged
5	„	24	„
6	„	25	pale red
7	„	26	„
8	„	27	„
9	„	28	„
10	„	29	lighter, yellowish
11	„	30	pale red
12	„	31	almost unchanged
13	„	32	„
14	„	33	pale yellow
15	white —	34	pale red
16	„	35	almost unchanged
17	light grey, almost white	36	pale yellow
18	„	37	„
19	„		

Even in England things do not appear to have been managed much better, for, according to my informant, the **only government-test for khaki-colored cottons** consisted in one hour's boiling in a 8 oz. of carbonate of soda per gallon, with replacement of the evaporated water. Such a test admits of the unrestricted use of coaltar colors, renders oxide-dyeing superfluous and will go far to explain the fact, that amongst a large quantity of cottoncloth imported from England via Hamburg during the China troubles there was rarely a piece of genuine khaki.

The official test for khaki-colored woollens, according to the "Dyer" is merely this: (1) 20 minutes boiling in weak carbonate of soda, (2) 30 minutes steeping in dilute cold peroxide of hydrogen. This is to serve as a criterion for fastness to washing and soap and to light and air!

It is evident that such tests cannot serve to distinguish between true metallic oxide dyeings and imitations. With a little practice however a simple immersion in acidulated chloride of lime solution suffices, in case an analytical method, such as the prussiate-test, should be declined, to give all the information desired, for there is not at present extant a single artificial khaki-dye, but what is either considerable altered or entirely bleached by chlorine.

As **to the value** of **the various tests suggested previously,** it may be gathered from the subjoined tables; the reagents employed being those specified on pages 27 and 30.

— Solidity-Test —

No. of pattern examined	Exposed from June 1 to July 15. 1902	Peroxide of sodium	Soda and soap	Acetic acid	Hydrochloric acid	Oxalic acid	Chloride of lime
1	unchanged	unchanged	unchanged	unchanged	unchanged	almost unchanged	unchanged
2	a little lighter	„	„	a little lighter	much lighter	almost bleached	a little lighter
3	a little darker towards grey	„	„	unchanged	lighter	„	unchanged
4	„	„	„	„	„	„	„
5	unchanged	„	„	„	unchanged	semi-bleached	„
6	„	„	„	„	„	„	„
7	„	„	„	„	„	a little better than 5	„
8	„	„	„	„	„	„	„
9	„	„	„	„	„	„	semi-bleached
10	„	„	„	„	„	„	„
11	„	„	„	„	little lighter	a little better than 5	„
12	„	„	„	„	„	semi-bleached	„
13	„	„	„	„	almost entirely bleached	almost entirely bleached	„
14	„	„	„	„	„	„	„
15	„	„	„	lighter	„	„	„
16	„	„	„	lighter	„	„	„
17	„	„	„	unchanged	unchanged	semi-bleached	„
18	„	„	„	„	„	„	„
19	„	„	„	a little lighter	almost entirely bleached	almost entirely bleached	„
20	darker towards grey	lighter and yellower	a little lighter	a little lighter	a little lighter	a little lighter	semi-bleached
21	unchanged	unchanged	unchanged	unchanged	almost entirely bleached	almost entirely bleached	a little lighter
22	„	„	„	„	„	„	„

That the severe ordeal, to which these dyeings were submitted, can and shall have a diagnostic value only, is selfevident, for practically fabrics will never be thus maltreated with acids and with chloride of lime. In any case, the fact becomes patent that, as regards fastness to light, metallic oxide colors are paramount; this, the creams and buffs, produced with iron salts, have proved long ago. In passing to khaki shades produced with organic dyestuffs, it may emphatically be said, that none of these attain the solidity to weather, which distinguishes **chrom-iron-khaki.** The defects of the latter, as regards its fitness for sewing, are however so serious, that its scope is exceedingly limited. In my opinion it should be used **for tent-cloth only.**

Shades produced with organic dyestuffs.

Berlin A. G.

No. 23.

Proportions employed:

sulphurbrown 2 G	1.5	percent.
calcined carbonate of soda .	5	„
sulphide of soda	2	„
salt (per gallon)	4	oz.

Raise the bath to the boil and add the soda; dissolve the dyestuff in some of the hot liquor under addition of sulphide of soda; pour into the bath and finally add the salt.

Wooden or iron vessels are to be used only.

Dye at the boil for 1 hour.

Berlin A. G.

No. **24**.

Dye with 1.5 percent of sulphurbrown G as with No. 23.

Farbenfabriken vorm. Fr. Bayer & Co., Elberfeld.

No. 25.

Proportions employed:

katigene yellowbrown G G . .	4.4 percent.
katigene green 2 B	0.067 „
sulphide of soda	2 „
calc. carbonate of soda . . .	2 „
sulphate of soda xals	7.5 „

Enter at the boil and continue boiling for 1 hour. Rinse and fix in a hot solution of

sulphate of copper . . .	4 percent.
acetic acid	5 „

Rinse and dry.—

Farbenfabriken vorm. Fr. Bayer & Co., Elberfeld.

No. 26.

(Matching the plumage of the skylark — alauda arvensis.)

Proportions employed:

			Additions required for standing bath:
katigene yellowbrown . . .	7	percent	4.7 percent
katigene blackbrown . . .	6	„	4 „
sulphide of soda	4	„	1 „
calc. carb. of soda	5	„	1 „
sulphate of soda xals. . .	30	„	3 „

Enter at the boil and continue boiling for 1 hour.

Rinse and fix for 30 minutes in a hot solution of:

sulphate of copper . . . . . 3 percent

acetic acid . . . . 5 „

Rinse and dry.—

Farbenfabriken vorm. Fr. Bayer & Co. Elberfeld.

No. 27.
("Fieldgrey".)

Proportions employed:			Additions required for standing bath:	
katigene yellowbrown GG. .	4.9	percent	3.3	percent
katigene blackbrown N. . .	2.1	„	1.4	„
katigeneblack SW.	1.75	„	1.2	„
katigene green 2 B.	0.7	„	0.47	„
sulphide of soda	4	„	1.5	„
calc. carb. soda	5	„	1	„
sulphate of soda xals. . . .	30	„	3	„

Enter at boil and continue boiling for 1 hour.

Rinse well and fix per 30 minutes in a hot bath of:

sulphate of copper . . . . . 3 percent
acetic acid . . . . 3 „

Rinse and dry.

Farbwerk Mühlheim vorm. A. Leonhardt & Co.

Proportions employed:

pyrolbronce G. . .	3.3	percent
pyrolbrown G. . .	0.3	,,
pyrolblack B. conc.	0.04	,,

Dissolve the dyestuff in 25 times its weight of boiling water and twice its weight of sulphide of soda; pour into the boiling bath and hereafter add 10 percent of calc. carb. of soda.

Dyeing is best done in the jigger. The goods are well wetted out before entering and run for one-half hour near the boil; continue for another hour, whilst adding 20 percent of salt in 3 to 4 lots

Finally rinse and dry.

No. 28.

Farbwerke vorm. Meister, Lucius & Brüning, Höchst o. M.

No. 29.

Proportions employed:

thiogenebrown G.	1.5	percent
thiogenebrown R.	1	,,
calc. carb. soda	2	,,
sulphate soda	20	,,

Enter boiling, continue for 1 hour, rinse well and fix in a hot bath of

bichromate of potash	2	percent
sulphate of copper	2	,,
acetic acid . . .	2	,,

Rinse and dry.—

Society for Chemical Industry, Basle.

Proportions employed:

$1\frac{1}{4}$ percent by weight of cloth of the following mixture

pyrogene catechu 2 G.	50	parts
pyrogenebrown G. .	**22**	„
pyrogeneyellow M. .	**28**	„

After dyeing fix by working for 20 minutes at 70—80° C. in a bath containing

sulphate of copper	1	percent by weight of cloth
acetic acid . .	1	„ „ „ „ „

No. 30.

Kalle & Co., Biebrich o. Rh.

No. 31.

Directions;

(a). Enter into a boiling bath of

thionbrown G. . .	0.6	percent
sulphanilinbrown 4B.	0.3	„
carbonate of soda .	10	„
sulphide of soda .	5	„

Work for 15 minutes at the boil and continue for ³|₄ hour, whilst cooling slowly.

(b). Rinse.

(c). Dry.

P.S. When working from a standing bath, lesser quantities of dyestuff etc. are employed.

Leopold Cassella & Co., Frankfort o. M.

Directions:

(a). Dye at the boil with:

immedial darkbrown A. 0.4 percent
immedial yellow D. . . 1.3 „
immedial black NG. . . 0.06 „

with the addition of

sulphide of soda	2/3 oz.	per gallon
caustic soda 76.6° Tw. . .	3 drams	
salt	1 oz.	

(b). Fix at the boil in a separate bath of:

sulphate of copper . . . . . 2 percent
bichromate of potash . . . . 1 „
acetic acid . . . . 3 „

(c). Rinse.

(d). Dry.

P. S. The above quantities are actually consumed.

No. 32.

Levinstein Ltd., Manchester.

No. 33.

Proportions employed:

thionolbrown G.	1	percent
dianolyellow B.	0.12	„
thionolblack G. T. X. . . .	0.08	„
calc. carb. soda	3	„
sulphide of soda	4	„
salt	10	„

Enter at the boil, continue for 1 hour, rinse and dry.

Read Holliday & Sons Ltd. Huddersfield.

Proportions employed:

amidazol-gambier	2,5	percent
calc. carb. soda	5	„
salt	10	„

Enter cold, raise to the boil within 30 minutes, boil another 30 minutes, rinse and dry.

No. 34.

Read Holliday & Sons Ltd. Huddersfield.

No. 35.

Proportion employed:

titan orange . . . . 0.6 percent
ingrain black 2 B. . . . . 0.04 „
ingrain black R. . . . . 0.05 „
salt . . . . 10 „

Enter cold, raise to the boil within 30 minutes, boil another 30 minutes, rinse and dry. —

Société Anonyme des Matières Colorantes etc. de St. Denis, Paris.

Dye at the boil with

cachou de Laval	1	percent
thiocatechine F.	0.5	„
carbonate of soda	10	„
salt	10	„

Rinse.—Fix in a boiling bath of

bichromate of potash	2.5	„
sulphate of copper	2.5	„
sulphuric acid	5	„

Rinse and dry.—

No. 36

Aniline color works vorm. J. R. Geigy, Basle.

No. 37.

Dye for 1 hour in the jigger with

eclipse-brown 3 G. . . . . 1 percent
eclipse-black B. . . . . 0.12 „
carbonate of soda . . . . 3 „
sulphide of soda . . . . 3 „
sulphate of soda (per gallon) 2/3 oz.

Hereafter fix with

bichromate of potash . . . . . 1 percent
sulphate of copper . . . . . . 1 „
acetic acid . . . . 2 „

The subjoined tables show the behaviour of the preceeding **khaki-imitations** if submitted to the **same tests,** as previously applied to metallic oxide khaki.

No. of pattern treated	40 days exposure	peroxide of sodium	soap and soda	acetic acid	hydro-chloric acid	oxalic acid	chloride of lime
No. 23	light grey	almost unchanged	almost unchanged	almost unchanged	almost unchanged	almost unchanged	dirty cream
24	light grey	"	"	"	"	"	"
25	much paler	"	"	"	"	"	"
26	almost unchanged	"	"	"	"	"	"
27	"	"	"	"	"	"	dirty light grey
28	light grey	"	"	"	"	"	dirty cream
29	a little lighter	a little lighter	"	"	"	"	remaining shade like No. 30
30	much paler	almost unchanged	"	"	"	"	dirty reddish yellow
31	a little lighter than 25	"	"	"	"	"	dirty cream
32	a little lighter than 31	"	little lighter	"	"	"	like 30
33	a little lighter	"	almost unchanged	"	"	"	a little brighter than 36
34	as 31	"	"	"	"	"	brown cream-color
35	a little lighter than 25	"	little lighter	"	"	"	orange color
36	greenish grey	"	almost unchanged	"	"	"	dirty sulphur-yellow
37	almost unchanged	"	"	"	"	"	light cream

Khaki in calicoprinting.

A.

Metallic oxide khaki.

Amongst the English patents relating to khaki, there is one, which also makes provision for **figured effects,** namely patent **No. 7888 A. D. 1899** granted **to F. J. Horrocks.** That part of the specification, relating to plain shades exclusively, has already been quoted on page 41. Further on the patentee expresses himself as follows:

—„If I desire to figure or pattern the fabrics, I engrave the surface of the top-squeezing-roller with the desired pattern, and when squeezing out the excess of liquor, the parts under the figures, if sunk below the surface of the roller, then contain an excess of liquor, while those parts, which have been pressed upon by the unremoved surface of the engraved roller, contain less of the liquor, and when the fabric is passed through the alkaline solution, these figures or parts, which have an excess of liquor, are much darker, than the ground work, out of which more of the liquor has been squeezed.

This figuring can only be done, when the fabric goes from one liquor (the mixed solutions of oxide of copper and of iron) direct to the other (the alkaline solution).“ —

Whatever be the actual value of such effects, they could doubtlessly be produced in a much safer manner and much neater, by first padding the cloth in a weak mordant, drying, printing with a suitable paste and hereafter passing it through the necessary developing and fixing liquors.—

For a direct khakiprint Lauber, in the "Monatsschrift für Färber und Drucker", No. 4, 1900 gives verbatim the following formula:

pyrolignite of iron . . .	2,5 gallons	
chloride of manganese .	5,5 lbs.	boil
green copperas	5,5 „	together,
sugar of lead	11 „	cool and
starch	3 „	add
british gum	1,5 „	
nitrate of iron 67° Tw.	12 oz.	

Print, dry and pass through alkali.

This produces a medium cutch; by increasing or reducing the proportion of manganese and iron salts, the shade may be varied at will.—

—As regards **discharge-work** on **true chrome-iron khaki**, the translator begs have to say this much: White patterns are readily produced by printing with thickened limejuice, if necessary strengthened with citric acid, passing through the Mather-Platt and washing. The color has to be considerably stronger, than what is currently employed for the old style of black and white; an acidity equal to 1,5 lbs. of citric acid per gallon should however suffice in all cases.

If an adequate amount of citrate of soda be added to such a white, it will be capable of resisting a vanadium or copper-anilineblack, so that white patterns on khaki under black covers may be produced.

If on the other hand powdered yellow prussiate be added to a discharge white of increased acidity, a bright pattern of prussian blue will result. Steaming has to occuppy several minutes; the print is then rinsed in hot water, in such a manner, that the air may further develop and fix the blue.

During the summer of 1900, there was a passing demand for bright red spots, stripes and other simple figures on khaki-twills for dressgoods. This effect is got without particular difficulty, by grounding the dyed fabric in the usual naphthol-oil-prepare, printing in a diazored, adequately acidulated with citric acid, ageing in the Mather-Platt, rinsing and soaping.—

B.

Khaki with artificial organic dyestuffs.

Apart from patterns No. 38—45, formulae for the production whereof are attached, the shades sampled as No. 23 and No. 24 and dyed with sulphurbrown G. and 2 G. of the Berlin A. G. may be discharged with the following printing-paste:

chlorate of soda 77° Tw. (being a solution of 53 oz of chlorate in 6 gills of water)	$2\frac{1}{4}$ lb.	boil, cool and add
british gum	$1\frac{1}{4}$ lb.	

powdered red prussiate . . . . . . $1\frac{1}{2}$ oz.
citric acid . . . . $3\frac{1}{2}$ oz.

Steam 1—3 minutes, wash and soap.

Discharge effects may also be produced with **tinpaste** on khakis, dyed with certain **"diaminecolors" (Cassella),** the solidity wherof to light may first be improved by aftertreatment with copper and bichromate. A shade, matching pattern No. 1 can for example be got with

diaminecatechine B. . . .	0.1	percent
diamineyellow N. powder	0.5	,,
diamineblackblue B. . .	0.02	,,

Dye as usual, adding pro gallon of bath soap 1 dram, soda 1 dram and phosphate of soda $\frac{2}{3}$ oz.—

A similar shade is got with

diaminecatechine B. .	0.09	percent
diamineyellow N. powder	0.45	,,
diamineblackblue . . .	0.01	,,

Dye as before and hereafter treat at the boil with: bichromate of potash 1 percent, sulphate of copper 1 percent, acetic acid 1 percent.—

Discharge-whites for these shades are the following:

(1). Ferrocyanide of tin discharge.

wheatstarch	4 oz.	boil together;
water	3 gills	
white dextrine	1 lb.	
citric acid	4 oz.	

add whilst warm

tincrystals	2 lbs.		

Stir well, cool and add

red prussiate	1 lb.	dissolved separately and cooled
water	1 gill	
gumsolution (1:1)	2 1/2 gills	

(2). Acetate of tin discharge.

wheatstarch	1 1/4 lb.	boil together;
acetate of tin 28.4° Tw.	4 gills	
dextrine	20 oz.	
citric acid	3 oz.	

add whilst warm

tincrystals . . . . . 1 1/2 lb;

cool and add

acetate of soda . . . 5—7 oz.

A little alkaliblue may be added to either color, to improve the white. Steam in the Mather-Platt and wash.

Badische Anilin- und Soda-Fabrik Ludwigshafen o. Rh.

Dyed with

kryogenebrown G. .	0.9	percent
kryogenebrown . .	6.35	,,

Discharge white:

chinaclay	1 lb.	make into a paste and add
water	$1^1/_4$ gill	
gumsolution (1:1) . . .	5 gills	
chlorate of soda . . .	2 lbs.	

Heat together, cool and add:

powdered tartaric acid .	$1^1/_4$ lb.
,, red prussiate .	3 oz.

Print, dry, age 7 minutes in the Mather-Platt, rinse and soap slightly.

No. 38.

Farbenfabriken vorm. Fr. Bayer & Co., Elberfeld.

No. 39.

Pad in Koechlin's alkaline-chromemordant, wash, dry and print with:

gum	2	lbs.
water	5	gills
crushed citric acid .	$1\frac{1}{2}$	lbs.
crushed tartaric acid	1	lb.

Pass through the Mather-Platt, wash and dye for 1 hour at the boil with

anthracenebrowu G.G. paste	1.5	percent
alizarine yellow R. paste	0.625	„
alizarine-viridine F.F. paste	0.125	„
acetic acid 9^0 Tw. . .	10	„
glue solution pro beck .	1	quart.

Wash thoroughly and dry.—

Farbenfabriken vorm. Fr. Bayer & Co., Elberfeld.

Dye the bleached cloth at the boil with

chrysophenine	0.25 percent.
benzochromebrown B.	0.1 „
benzochromebrown G.	0.1 „
direct-deep-black R. W. . . .	0.02 „
sulphate of soda	10 „
calc. carb. soda	2 „

Wash well, dry, print, steam for 10 minutes, wash and dry.

No. 40.

Printing color: 1 part by weight of "discharge paste*)"
2 parts by weight of tragacanth-starch-thickening.

*) "Discharge paste":

wheatstarch.	$1^1/_4$ lb.	boil together;
acetate of tin 32.4° Tw.	1 quart	
gumsolution 1 : 1	1 pint	
water.	1 quart	
add whilst hot tincrystals	1 lb.	
when lukewarm add		
crushed citric acid	4 oz.	

Farbwerk Mühlheim vorm. Leonhardt & Co., Mühlheim o. M.

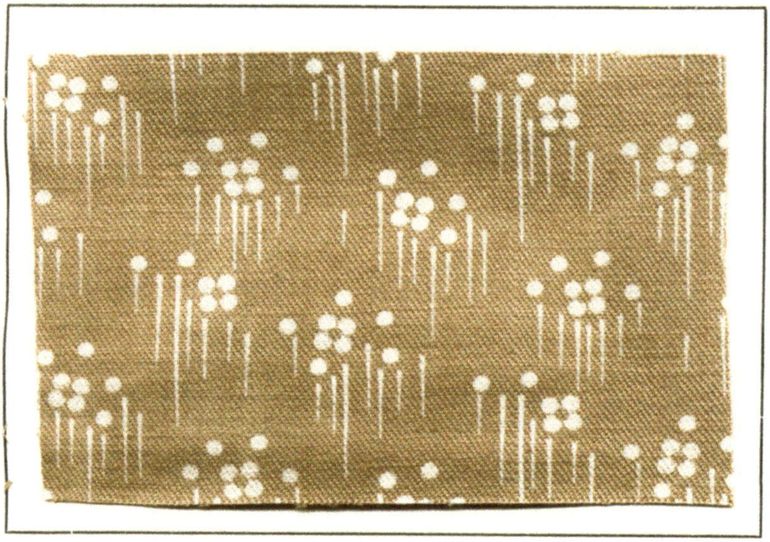

No. 41.

Proportions employed for jigg-dyeing:

pyrolbronce G.	3.3 percent	
pyrolbrown G.	**0.3**	„
pyrolblack B. conc. . .	0.04	„

Dissolve the dyestuff in double the weight of sulphide of soda and 25 times the weight of boiling water, pour into the boiling bath and hereafter add 10 percent of calcined carbonate of soda.

Enter the wet goods and run at the boil for 1/2 hour; then run for another half-hour, whilst adding 20 percent of salt in 3—4 lots.

Rinse and dry.—

Discharge white:

british gum	110	parts by weight	boil together
chlorate of soda 76.6° Tw.	126	„ „ „	and add
chinaclay	62	„ „ „	
cool and add			
citric acid solution (1 : 1)	36	„ „ „	
chlorate of alumina 53° Tw.	35	„ „ „	
crushed red prussiate . . .	7	„ „ „	

Print, dry, pass through the Mather-Platt, soap hot with 1/4 oz. of soap and 1/4 oz. of calcined carbonate of soda per gallon, rinse and dry.—

Farbwerke vorm. Meister, Lucius & Brüning, Hoechst a. M.

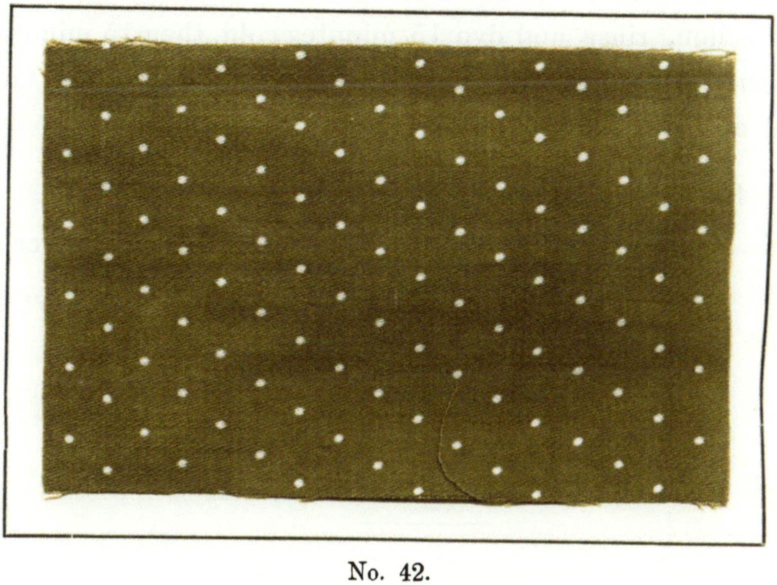

No. 42.

1st operation:

Pad with	"chromemordant G. A. II"	$1\frac{1}{4}$	lb.
	british gum paste	$1\frac{1}{4}$	lb.
	glycerine	8	oz.
	water	—	
		to make $2\frac{1}{2}$	gallons

2nd operation:

Dry and print with

british gum		2	lbs.
citrate of ammonia	48.2° Tw.	1	pint
sulphite of potash .	90° Tw.	$1\frac{1}{4}$	lb.
water		3	pints

Steam for $\frac{1}{2}$ hour.

3rd operation:

Wash for 15 minutes in 2 oz. of carbonate of soda per gallon, rinse and dye 15 minutes cold, then 15 minutes warm, then 15 minutes at 60° C; slowly raise to the boil and continue boiling for another half-hour.

Dyebath:

alizarinebrown paste	1.9	percent
alizarineyellow G.G. paste	1.14	,,
alizarinegreen S. paste	0.56	

water: 50 times the weight of cloth,
acetic acid: 2 drams per gallon.

Farbwerke vorm. Meister, Lucius & Brüning, Hoechst o. M.

No. 43.

Twocolor print.

Blotch.

phosphine O.	1	gram
leather-yellow O.	0.6	,,
methylenegreen extrayellow D.	0.3	,,
acetine	20	,,
glycerin	20	,,
acetic acid 12° Tw.	100	,,
gumsolution (1 : 1).	600	,,
tartaric acid	1	,,
acetic tannin solution (1 : 10)	80	,,
water	177	,,
	1	kilo.

Spray:

methyleneheliotrope O. . . .	1	gram
methyleneviolet 3 R.A. . . .	4	,,
acetic acid 12° Tw.	100	,,
acetine	20	,,
glycerin	30	,,
starchpaste	600	,,
tartaric acid	2.5	,,
acetic tannin solution (1 : 10)	25	,,
water	218	,,
	1	kilo

Steam for 1 hour, pass through tartar emetic, rinse and soap.

Leopold Cassella & Co., Frankfort o. M.

No. 44.

Dye boiling on the jigger with:

thioflavine S . . .	2.3 %	
diaminecatechine 3 G	1.3 ,,	
diaminecatechine B .	0.4 ,,	
calc. carb. soda . . .	1/8 oz.	per gallon.
calc. sulph. soda . .	2 oz.	per gallon.

Rinse, dry and print as follows:

Yellow: This color, due to the thioflavine used for the bottom, requires merely a tin-discharge-paste, made by boiling together:

wheatstarch	4	oz.
acetate of tin 28° Tw. .	1	pint
white dextrine . . .	8	oz.
citric acid	1,5	,,

Green: In 1/2 gill acetic acid 9° Tw. and 1/2 gill of water dissolve 2/3 oz. of thioflavine T and 1/3 oz. of brilliant-green,—thicken with 1/2 gill of gum solution,—add 1/2 gill of acetic-tannin-solution (1 : 1) and work the whole into 1 1/2 gill of the above tin-discharge-paste.

Darkbrown: In 1/2 gill acetic acid 9° Tw., 1/16 gill of acetine and 1/3 gill of water dissolve 12 drams of paraphosphine R, 2 drs. of safranine S 150 and 1 1/2 drs. of crystalviolet 5 B. Add 2 1/2 drs. of tartaric acid in 1/8 gill of water. Work the whole into 1 1/2 gills of tragacanth-starch-paste and finally add 1/2 gill of acetic-tannin-solution (1 : 1).

Lightbrown: Darkbrown, reduced 6 times. —

After printing steam for $\frac{1}{2}$—$\frac{3}{4}$ hour, pass through tartar-emetic and wash.

Leopold Cassella & Co., Frankfort o. M.

No. 45.

Print with anilineblack and hereafter dye with

diamine-fast-yellow B. . . .	0.7 percent		
diamine catechine B.	0.5 „		
calc. carb. soda	$\frac{1}{4}$ oz.	}	per
calc. sulphate of soda . . .	$\frac{2}{3}$ oz.	}	gallon

Khaki on linen and jute.

In so far as the dyeing of linen and jute is carried out on the same lines as that of cotton, the methods already described for the production of chrome-iron khaki are also available in this instance.

It will however be found that, even when the same particulars are adhered to, the dyeings on linen, though they satisfactorily resist all other reagents, are less fast to washing (that is one-hours boiling with 5 drams of soda and 12 drams of soap per quart), than those on cotton. This has no doubt to be ascribed to the closer texture and lesser permeability of the linenfibre; the metallic oxides are percipitated in part on the surface and are consequently more liable to be removed, particularly by friction.

It has sometimes been said that chrome-iron khaki is altogether unsuitable for linen; this is going too far, but nevertheless certain precautions are indispensable, in order to attain good results. Before all, the goods must be thoroughly wetted out; they should for this purpose be given at least 10 – 12 ends at the boil in the jigger and hereafter be washed. They are then, without previous drying, ready for dyeing. Repeated passages through the mordant and the alkali, all without intermediate drying, are required, and each turn through soda should be followed by a thorough washing, by preference in a machine fitted with beaters, so that all the loose oxide may be removed before reentering the mordant. In this manner shades fast to washing can be got.

As linen-goods are at the outset more difficult to sew than cotton ones, the particular drawbacks attached to oxide-dyeings become all the more pronounced in the present instance.

As regards the production of khakishades by combining wood colors with metals, what has been said on pages 58—60 holds equally good with linen.

Herzfeld, in "Die Praxis der Färberei" (M. Krayn, Berlin) gives the following recipes:

I. Light mode (for 55 lbs. of linencloth).

Work at 30° C. in a bath containing 2 oz. of tannic acid and sadden in the same bath with $5\frac{1}{2}$ lbs. of pyrolignite of iron.

II. Medium mode (for 55 lbs. of linencloth).

Set the bath with catechu 2 lbs., tannic acid 5 oz. and quercitron-extract (solid) 2 oz.; work at 40° C. and sadden in the same bath with $4\frac{1}{2}$ lbs. of nitrate of iron.

III. Dark mode (for 55 lbs. of line cloth).

Set the bath with catechu $3\frac{1}{4}$ lbs., tannic acid 8 oz., quercitron-extract 4 oz.; work at 40° C. and sadden with $6\frac{1}{2}$ lbs. of nitrate of iron.

IV. Olive mode (for 55 lbs.).

Set the bath with catechu $3\frac{1}{4}$ lbs., quercitron-extract $3\frac{1}{4}$ lbs. and tannic acid 4 oz.; work at 40° C. and sadden in the same bath with $6\frac{1}{2}$ lbs. of nitrate of iron.

V Drab. . . . . (for 55 lbs.).

Set the bath with catechu $2\frac{1}{4}$ lbs., sulphate of copper 5 oz. and logwood-extract 10 oz.; work at 40° C.; sadden in the same bath with 12 oz. of green copperas and hereafter develop in a fresh bath at 60° C. with 1 lb. of bichromate of potash.—

Khakishades on linen, produced with artificial organic coloring matters are shown on the subjoined patterns No. 46—51.

Berlin A. G.

No. 46.

Proportions employed:

sulphurbrown G. . . .	0.75	percent.
columbiayellow	0.12	„
carb. soda	5	„
sulphide of soda . . .	3	„
sulphate of soda . . .	5	„

Work for 1 hour at 80° C., rinse and dry.

Farbenfabriken vorm. Fr. Bayer & Co., Elberfeld.

No. 47.

Dye at the boil with:

katigenechromebrown 5 G. . . .	0.6 percent.
katigeneyellowbrown G. G. . . .	1 ,,
katigeneolive G. N.	0.05 ,,
calc. carb. soda.	2 ,,
sulphide of soda	1.5 ,,
calc. sulphate of soda	5 ,,
turkeyred oil	3 ,,

Hereafter fix at 80° C. in separate bath containing:

bichromate of potash . . .	3 percent.
sulphate of copper	3 ,,
acetic acid	3 ,,

The above quantities are actually consumed, the bath being practically exhausted.

Farbenfabriken vorm. Fr. Bayer & Co., Elberfeld.

Dye at the boil with:

katigenechromebrown 5 G. . . .	0.75 percent.
katigeneyellowbrown G. G. . . .	1 ,,
katigeneolive G. N.	0.70 ,,
calc. carb. soda	2 ,,
sulphide of soda	2.5 ,,
sulphate of soda	5 ,,
turkeyred oil	3 ,,

Hereafter treat in a separate bath with:

bichromate of potash . . .	3 percent.
sulphate of copper	3 ,,
acetic acid	3 ,,

The above quantities are actually consumed, the bath being practically exhausted. —

No. 48.

Farbwerke vorm. Meister, Lucius & Brüning, Hoechst o. M.

No. 49.

Dye at the boil with:

thiogenebrown 2 G	**0.6 percent.**
oxydianilyellow O.	**0.1 „**
calc. carb. soda	**2 „**
salt	**5 „**

Fix in a hot bath containing:

sulphate of copper.	0.75 percent.
bichromate of potash	0.75 „
acetic acid	2 „

Leopold Cassella & Co., Frankfort o. M.

Dye at the boil with:

immedialbrown R. R.	0.6 percent.	
immedialdarkbrown A.	0.5 ,,	
immedialyellow D.	1.1 ,,	
carbonate of soda	1/2 oz.	per gallon.
sulphide of soda	1 oz.	per gallon.
turkeyred oil	1/4 oz.	per gallon.
sulphate of soda	2/3 to 1 1/2 oz.	per gallon.

Hereafter treat in a separate boiling bath with:

bichromate of potash	1.5 percent.
sulphate of copper	1.5 ,,
acetic acid	3 ,,

Rinse and dry.

The above quantities of dyestuff are actually consumed and as much more is required in setting the bath, as corresponds with its volume.—

No. 50.

Leopold Cassella & Co., Frankfort o. M.

No. 51.

"Fieldgrey".

Dyed at the boil with:

immedialolive B.	0.5 percent.	
immedialdarkbrown A.	0.4 ,,	
immedialblack N. G.	1.15 ,,	
carbonate of soda.	$\frac{1}{2}$ oz.	per gallon.
sulphide of soda	1 oz.	
turkeyredoil	$\frac{1}{4}$ oz.	
sulphate of soda	$\frac{2}{3}$ to $1\frac{1}{2}$ oz.	

Fix at the boil in a separate bath with:

bichromate of potash	1.5 percent.
sulphate of copper	1.5 ,,
acetic acid	3 ,,

The above dyestuff quantities are actually taken up and as much more is required in setting the bath, as corresponds with its volume.—

Khaki on wool and unions.

Though solid khakishades can be produced on wool by ordinary methods, metallic oxide colorings have nevertheless been produced even on this material.

There are on record 2 English patents, relating to this subject.

The one, granted to **Frederick Albert Gatty, No. 9286 A. D. 1886,** rests on the following specification:

—"My improvements relate to the fixing of oxide of chromium, or oxide of chromium combined with a certain proportion of chromic acid, upon wool or other animal fibres in their raw or manufactured state.

In place of boiling the wool or other animal fibre with chromate of soda or chromate of potash and sulphuric acid or other acid, as generally practised in dyeing wool, I apply the chromate of soda or chromate of potash with the requisite quantity of sulphuric acid to set the chromic acid free at the ordinary temperature of the atmosphere or only slightly warmed.

After the wool has been worked from one to two hours in the acid chromate solution, it is washed and then subjected to the action of boiling water or steam to deoxidise the chromic acid fixed upon the wool and thus form oxide of chromium. The wool or other animal fibres, treated as described above, is then ready for dyeing in the usual way.

For certain colours some chromic acid in combination with the oxide of chromium is required. For such colours the wool, prepared with the acid chromate, is exposed for a shorter time to the action of boiling water or steam, than the time required for the decomposition of the whole of the chromic acid.

In certain cases other substances, such as salts of alumina, iron or tin may be fixed upon the wool in combination with the oxide of chromium for such colours, as require them either as mordants for dyeing, or as self-colours.

And, in order that my invention may be fully understood and readily carried into effect, I will proceed to describe in detail a process for carrying out the same, which process I find in practice gives good results.

The wool or other animal fibres, hereinafter referred to as wool only, which is to be dyed according to my invention, may be either in its raw state or spun or made into cloth, and is first cleaned and prepared in the usual way for dyeing. The wool, thus freed from oil and other impurities, is then put into a bath prepared as follows.

For each two hundred pounds weight of wool, woollen yarn or cloth, I dissolve in three to four hundred gallons of cold or nearly cold water from four to twelve pounds of bichromate of potash or bichromate of soda, the proportion of bichromate being varied according to the quantity of oxide of chromium desired to be fixed upon the wool, and to this solution I add from one half to three quarters of a pound of sulphuric acid for every pound of bichromate of potash or bichromate of soda dissolved.

If, instead of the bichromates, the chromates of potash or soda be used, the quantity of sulphuric acid must be increased by, say, one half.

After the wool has been worked or moved about in this solution, which I designate the chrome bath, for from one to two hours or longer if required, it is taken out and slightly washed and is then boiled in water for from two to three hours, in order to more or less deoxidise the chromic acid fixed upon the wool.

In order to hasten the deoxidising action of the boiling water, one to two pounds of sulphuric acid of commerce may be added to every hundred gallons of water employed. The addition of a certain quantity of a vegetable acid, such as oxalic or tartaric acid, or vegetable substance, such as sugar or starch, will hasten the deoxidation still more, but I prefer the use of boiling water, either with or without sulphuric acid.

I have stated here, and also in describing the composition of the chrome bath, that I use sulphuric acid. It will be obvious, however, to all chemists, that other mineral acids might be employed instead. I prefer, however, to use sulphuric acid, as being the best for this purpose and the most economical.

When the wool has been sufficiently deoxidised, and this can be easily seen by the change of colour from a yellow to a green or greenish shade, it is taken out, and if only water has been used, it is at once ready for dyeing in logwood, or other dyewood or alizarine or other colouring matter, in the usual way. If, however, sulphuric acid has been used in the boiling operation, the wool must be washed in water, before being dyed.

In place of boiling the wool in water after the chrome bath, as hereinbefore described, it may be steamed for from two to three hours, which operation will also deoxidise the chromic acid upon the wool like the boiling in water.

In cases, where metallic salts are used in combination with the oxide of chromium, if the substance required be iron, I put a certain quantity of a salt of iron into the chrome bath, and for this purpose I prefer to use sulphate of peroxide of iron, which answers well, but other salts of iron may also be used and good results obtained. If alumina be required, I add alum, and if tin be required, I add oxymuriate of tin to the chrome bath.

The oxide of chromium alone, and oxide of chromium combined with oxide of iron or chromate of iron, being colours in themselves, for such colours the wool need only be washed after the boiling in water or the steaming operation and then dried."—

The invention, protected by the subsequent patent **No. 29150 A. D. 1896** by **F. A. Gatty & Co. Ltd.** and **V. H. Gatty,** is described as follows:

"In carrying the improvements into effect, we take a soluble salt of chromium $\frac{\text{and}}{\text{or}}$ a soluble salt of iron and to their solutions, or a mixture of them, we add sufficient alkali to render the solutions basic.

For example, we take a solution of any soluble salt of iron, such as the sulphate or acetate, of a strength of about twenty degrees of Twaddles hydrometer, and mix therewith an equal amount of a solution of any soluble salt of chromium, such as the chloride or bisulphite of the same strength, and add thereto sufficient of a solution of carbonate of soda or other suitable alkali, to render the solutions basic.

In performing this, we take the solutions of the normal salts and add thereto a strong solution of carbonate of soda or other suitable alkali gradually and with constant stirring, until the point is reached, where the precipitate, which is at first formed, is no longer redissolved. We find that, in the case of a mixture of equal parts of the solutions of sulphate of iron and chloride of chromium at twenty degrees of Twaddle's hydrometer, an addition of three pounds of carbonate of soda crystals to each four gallons of the mixed solutions is necessary to produce the required result.

Where other salts or other proportions are used, the requisite amount of alkali must be determined experimentally.

Or we take basic solutions of the two salts produced in any other suitable and convenient way, such for instance, as by adding to the solutions of the normal salts as much of the hydroxides of the respective metals, as the solutions will dissolve.

We vary the strength and relative proportions of the two solutions of iron and chromium salts according to the colour required. In a bath thus formed we immerse the wool or woollen cloth or yarn, to be dyed, until saturated, and wring or squeeze out the surplus liquor. The wool absorbs the basic oxides or salts from their solution, a longer immersion giving a fuller shade and vice versa, but we prefer to regulate the shade as well as the colour by the strength and proportions of the iron and chromium salts in the liquor used. The wool (or woollen goods) may then be packed together and left wet for several hours (suitable precautions being taken to prevent uneven drying), to complete the absorption of the basic oxides or salts, or the wool may be subjected to the action of steam or ammonia gas, or of steam and ammonia gas together, to complete the fixation of the basic oxides or salts, or it may simply be dried in any suitable way after wringing out the surplus liquor, or we may use any of these methods in combination. The wool is then washed, to remove any surplus or unfixed colouring matter, and, after drying, is then ready for use.

In like manner we may employ a single solution of either a soluble salt of iron or of chromium, and after rendering it basic with an alkali or otherwise, proceed to dye wool or woollen cloth or yarn with it, as described for the mixed basic solutions."—

—According to Gärdner, the desired result is attained in a manner more simple, by first boiling the wool with bichromate and tartaric acid, and hereafter again with sulphate of iron merely.—

In the production of khaki-shades on wool, wood-dyes have been almost entirely abandoned in favour of artificial coloring matters, which yield results, answering all requirements for fastness to milling and light.

The following formula from the Leipziger Färber-Zeitung may nevertheless be quoted:

Work for 4 hours at about 80° C. in a bath made of catechu 10 lbs. and fustic 2 lbs.; hereafter add 1 lb. of bichromate, boil for half-an-hour, lift, rinse and dry.

Berlin A. G.

No. 52.

Enter the wool at 60° C. in a bath containing 3 percent of "metachrome-mordant", raise to the boil and boil for half-an-hour. Hereafter add 1 percent of ammonia and the following dyestuffs:

metachromebrown B.	0.1 percent.
metachromeorange R.	0.15 „
indocyanine B.	0.035 „
metachromeyollow R.	0.005 „

Boil for 2 hours and rinse.—

Berlin A. G.

First dye the wool, as shown with pattern No. 52, with:

metachromebrown B.	0.06 percent	by weight of wool.
metachromeorange R.	0.075 ,,	
indocyanine B.	0.017 ,,	
metachromemordant	1.5 ,,	

After rinsing, fill up in a separate cold bath with the following cotton-colors:

micadoorange 4 R. O.	0.15 percent	by weight of cotton.
curcumine S.	0.5 ,,	
columbiablue G.	0.05 ,,	
calc. sulphate of soda . . .	10 ,,	

Work for 3/4 hour and rinse.

No. 53. (Wool and cotton.)

Farbenfabriken vorm. Fr. Bayer & Co., Elberfeld.

No. 54.

Mordant with:

bichromate of potash 3 percent
cream of tartar . . . 2.5 „

Dye with:

anthraceneyellow paste 3.3 percent
alizarineblueblack B. . . 0.02 „
anthracenebrown W. . . 0.15 „

Farbwerke vorm. Meister, Lucius & Brüning, Höchst o. M.

Dye with:

S. A. brown B. . . .	0.24	percent.
mordantyellow O . .	0.16	"
chromotrope F 4 B. .	0.03	"
sulphate of soda . .	20	"
sulphuric acid. . . .	3	"

Hereafter treat with:

bichromate of potash 1 percent.

No. 55.

Farbwerke vorm. Meister, Lucius & Brüning, Höchst o. M.

No. 56.

Dye with:

chromogene I.	0.7	percent
alizarineyellow G.G.W.	0.09	„
chromotrope F 4 B. . .	0.025	„
sulphate of soda . . .	20	„
sulphuric acid.	4	„

Hereafter develop with:

bichromate of potash	1	percent

Kalle & Co., Biebrich o. Rh.

Dye with:

naphtaminebrown 8 B. . . 0.07 percent
naphtamineyellow G. . . . 0.35 „
chrysophenine G. . . . . . 0.10 „
Biebrich patent black 4 AN. 0.02 „
orange II . . . . 0.07 „
sulphate of soda $1^1/_2$ oz. per gallon.

Enter at the boil and work for 1 hur.

No. 57.
(Half-wool)

Kalle & Co., Biebrich o. Rh.

No. 58.

Produced with:

salicinered G.	0.05	percent.
salicineyellow G. . . .	0.15	„
chromepatentblack NB.	0.13	„
sulphate of soda xals.	10	„
sulphuric acid	2	„

Enter just warm, raise to the boil within 20 minutes and boil for 45 minutes; hereafter add bichromate of soda 0.5 percent and boil for another half-hour.

Leopold Cassella & Co., Frankfort o. M.

Produced with:

anthracenechromebrown D.	0.3	percent.
anthraceneyellow BN. . . .	0.1	,,
sulphate of soda xals. . .	10	,,
sulphuric acid	1	,,

Enter just warm, raise to the boil in 30 to 40 minutes, continue for $1^1/_4$ hour and hereafter develop by 30 minutes boiling in the same bath with

bichromate of potash . . . 0.5 percent. —

No. 59.

Leopold Cassella & Co., Frankfort o. M.

No. 60.

Produced with:

anthracenechromebrown .	0.5	percent
anthraceneyellow BN. . .	0.18	,,
sulphate of soda xals. .	10	,,
sulphuric acid	1	,,

Enter just warm, raise to the boil in 30 to 40 minutes, continue for $1\frac{1}{4}$ hour, then add

bichromate of potash . . 0.6 percent.

and boil for another half-hour.

Leopold Cassella & Co., Frankfort o. M.

Produced with:

diamine-fast-yellow B.	0.28	percent.
diamineorange B.	0.25	„
diaminecatechine B.	0.14	„
diaminogene extra	0.09	„

sulphate of soda xals. 13 oz. per gallon.

Enter at 60 ° C., raise to the boil in 30 minutes, and continue boiling for $^3/_4$ hour.

No. 61.
(Half-wool).

Leopold Cassella & Co., Frankfort o. M.

No. 62.
(„Fieldgrey“ on wool.)

Produced with:

anthracenechromebrown D 0.55 percent.
anthraceneblue C. . . . . . 0.22 „
anthraceneyellow BN. . . 0.09 „
sulphate of soda xals. . . 10 „
sulphuric acid . . . . 1 „

Enter just warm, raise to the boil in 30—40 minutes, boil 1¼ hour; then add

bichromate of potash . . . 0.6 percent

and boil another half-hour.—

Khaki on silk and silkunions.

Khaki shades of satisfactory solidity are readily produced upon these materials with coaltar-colors, — metallic oxide or wood-colorings can therefore be dispensed with.

Kkaki shades can be got **on silk** under others with the following dyestuffs or rather dyestuff mixtures:

I.	azoflavine FF. azoocarmine G. paste nigrosine W.	B. A. & S. F.
II.	solidyellow G. azomagenta G. (or azogarnet S.) induline B. (or alizarine-saphiroleB.)	Farben-fabriken, Elberfeld.
III.	Domingo-chromeyellow. Domingo-chromeblack. Domingo-chromered.	Farbwerk Mühlheim (vorm. A. Leonhardt & Co.)
IV.	azoyellow (conc.). fast acid violet R. nigrosine A.	Farbwerke, Höchst.

Khaki shades or halfsilk may be produced under others with the following combinations:

I.	chrysophenine. Pluto orange benzofastred GL or L. benzofastblue BM or benzofastblack	Farbenfabriken Elberfeld.

II. oxydianilyellow O. dianilbrown B. D. dianilazurine shaded with: azoyellow conc. and patentblue V.	Farbwerke, Höchst.

Leopold Cassella & Co., Frankfort o. M.

No. 63.
(Silk.)

Produced with:

cyanol extra. . . 0.45 percent
orange extra. . . 1 „
indianyellow. . . 0.08 „

Kalle & Co., Biebrich o. Rh.

Produced with:

naphtaminebrown 8 B. . . . 0.225 percent.
naphtamineyellow G. . . . . 0.8 „
chrysophenine G. . . . . . . 0.15 „
Biebrich patentblack 4 A N. 0.02 „
sulphate of soda $1\frac{1}{2}$ oz. per gallon.

Enter at 90° C. and work for 1 hour at this temperature.

No. 64.
(Halfsilk.)

Conclusions.

The outcome of the preceeding pages may be summed up thus:

On cotton and linen no dyestuff produces shades, equalling those got with metallic oxides for fastness to atmospheric action.

The cheapest materials for chrome-iron khaki are chromealum, nitrate of iron and pyrolignite of iron.

The use of pure acetates of chromium and iron has no notable effect on the shade; the strength of the fabric however is influenced less favourably, than with the pyrolignites.

The proper rotation of operations in khaki dyeing is:

(1). Impregnation with metallic salts.
(2). Fixing with boiling soda.
(3). Washing und drying.
(4). Steaming for 1 hour at 15 lbs. pressure.
(5). Aftertreatment with boiling silicate.
(6). Washing and drying.

So-called "fieldgrey" shades cannot be produced with metallic oxides; wood colors are too fugitive to light; certain artificial coloring matters, on the contrary, yield shades, exceedingly fast to exposure, as demonstrated on cotton by pattern No. 27 and on linen by pattern No. 51.—

On animal fibre, khaki shades of satisfactory solidity may be got with artificial coloring matters; thus patterns No. 54, 56 and 61 stood 40 days exposure during autumn without perceptible change.—

A. W. Hayn's Erben, Berlin und Potsdam.

Bei Fragen zur Produktsicherheit wenden Sie sich bitte an:
If you have any questions regarding product safety,
please contact:

Walter de Gruyter GmbH
Genthiner Straße 13
10785 Berlin
productsafety@degruyterbrill.com